SPRINGER
LABORATORY

Bernd Wenclawiak (Ed.)

Analysis with Supercritical Fluids: Extraction and Chromatography

With 117 Figures

Springer-Verlag
Berlin Heidelberg New York London Paris
Tokyo Hong Kong Barcelona Budapest

Professor Dr. BERND WENCLAWIAK

Universität-GH Siegen
Fachbereich 8
Analytische Chemie I
Adolf-Reichwein-Straße
W-5900 Siegen, FRG

ISBN 3-540-55420-3 Springer-Verlag Berlin Heidelberg New York
ISBN 0-387-55420-3 Springer-Verlag New York Berlin Heidelberg

Library of Congress Cataloging-in-Publication Data. Wenclawiak, Bernd. Analysis with supercritical fluids: extraction and chromatography, Bernd Wenclawiak. p. cm. Includes bibliographical references and index. ISBN 3-540-55420-3 (Springer-Verlag Berlin Heidelberg New York: acid-free paper): DM 148.00. – ISBN 0-387-55420-3 (Springer-Verlag New York Berlin Heidelberg: acid-free paper) 1. Supercritical fluid chromatography. 2. Supercritical fluid extraction. I. Title. QD79.C45W46 1992 543'.089–dc20 92-18355

Typesetting: K+V Fotosatz GmbH, Beerfelden
52/3145-5 4 3 2 1 0 – Printed on acid-free paper

Preface

The use of supercritical fluids in analytical chemistry is still growing. More and more analysts are discovering the favorable advantages for a number of applications. Especially supercritical fluid extraction (SFE) has attracted a lot of interest in recent years due to its simplicity. Supercritical fluid chromatography (SFC) has become better established and the development of this technique has been accelerated by the many applications with capillary columns which have been published in the literature.

At first SFC equipment was based on instruments commonly used for liquid chromatography, and the first commercial instruments were derived from this technology. However, capillary columns can be much more easily interfaced to gas chromatography equipment especially to the detectors commonly used for GC.

Many stationary phases both for packed micro columns and capillary columns have been designed for SFC purposes extending this technology to LC and GC.

The most common fluid applied in SFC and SFE is carbon dioxide. The advantages of supercritical CO_2, such as having diffusivity like a gas and solvating power depending on temperature and pressure, are also valid for other fluids and modified fluids. Both properties are valuable for sample extraction and extraction selectivity.

The link between much engineering research work and analytical chemistry research is still weak. Much can be learned from both fields. As soon as more standard procedures will allow or ask for the use of supercritical fluids, especially carbon dioxide, many chemists will apply them. Instrument manufacturers are working hard to fulfill the expectations and requirements of many scientists. Supercritical fluid chromatography has become established as an alternative and the missing link between liquid chromatography and gas chromatography. What can be expected from SFE? It is conquering the analytical laboratory and is competing with standard procedures. As soon as automatic instrumentation with high sample capacity becomes available, it will become the standard procedure for sample preparation of many substances.

This book contains articles collected over a period of 1 year from various contributors whose efforts are outlined in the various chapters. I have to thank all the authors for their work and time to make this a sound undertaking.

My thanks go also to H. Steuber for typing and retyping many of the manuscripts and to D. Winkel for working on some figures, also to the publisher and to all of those who patiently supported my project.

Siegen, August 1992

BERND WENCLAWIAK

Contents

List of Contributors

ELIZABETH M. CALVEY
Food and Drug Administration
Division of Contaminants
Washington, DC 20204, USA

THOMAS A. DEAN
Department of Chemistry
Wayne State University
Detroit, MI 48202, USA

JOHN E. FRANCE
Northern Regional
Research Center
Agricultural Research Service
United States Department of
Agricultural
1815 N. University Street
Peoria, IL 61604, USA

STEVEN R. GOATES
Department of Chemistry
Brigham Young University
Provo, UT 84602, USA

J. TYGE GREIBROKK
Department of Chemistry
University of Oslo
P.B. 1033 Blindern
0315 Oslo 3, Norway

STEVEN B. HAWTHORNE
Energy and Environmental
Research Center
University of North Dakota
Grand Forks, ND 58202, USA

JERRY W. KING
Northern Regional
Research Center
Agricultural Research Service
United States Department of
Agricultural
1815 N. University Street
Peoria, IL 61604, USA

ERNST KLESPER
Lehrstuhl für
Makromolekulare Chemie
Technische Universität Aachen
Worringerweg 1
W-5100 Aachen, FRG

MILTON L. LEE
Department of Chemistry
Brigham Young University
Provo, UT 84602, USA

DAVID M. LUBMAN
Department of Chemistry
University of Michigan
Ann Arbor, MI 48109, USA

JOHN W. OUDSEMA
Department of Chemistry
Wayne State University
Detroit, MI 48202, USA

J. DAVID PINKSTON
The Procter & Gamble Company
Miami Valley Laboratories
P.O. Box 398707
Cincinnati, OH 45239, USA

Colin F. Poole
Department of Chemistry
Wayne State University
Detroit, MI 48202, USA

Salwa K. Poole
Department of Chemistry
Wayne State University
Detroit, MI 48202, USA

Michael Schleimer
Institut für Organische Chemie
der Universität Tübingen
Auf der Morgenstelle 18
W-7400 Tübingen, FRG

Franz P. Schmitz
Lehrstuhl für
Makromolekulare Chemie
Technische Universität Aachen
Worringerweg 1
W-5100 Aachen, FRG

Gerhard M. Schneider
Physikalisch, Chemisches
Laboratorium
der Universität Bochum
Abteilung Chemie
W-4630 Bochum 1, FRG

Volker Schurig
Institut für Organische Chemie
der Universität Tübingen
Auf der Morgenstelle 18
W-7400 Tübingen, FRG

Chung Hang Sin
Department of Chemistry
Northern Illinois University
DeKalb, IL 60115, USA

Larry T. Taylor
Virginia Polytechnic Institute
and State University
Department of Chemistry
Blacksburg, VA 24061-0212,
USA

B. Wenclawiak
Universität GH Siegen
Analytische Chemie I
Adolf-Reichwein-Straße
W-5900 Siegen, FRG

1 SFC and SFE: An Introduction for Novices

BERND WENCLAWIAK

Historically, supercritical fluid chromatography (SFC) was first demonstrated in 1962 [1]. The number of papers published each year during the following two decades did not encourage many researchers to investigate this technique [2–4]. It was not until 1981 after fused silica capillary column SFC was introduced that the number of reports steadily increased, levelling off in recent years [5].

Since the mid 1980s reports on analytical supercritical fluid extraction (SFE) have been published [6]. On- and off-line techniques are in use and much work on SFE has been done by engineers on a larger technical scale [7–9].

Before we can discuss any specific topic about supercritical fluid chromatography or extraction (SFC/SFE) we should remember what a supercritical fluid is. Looking at a single component pressure-temperature (pT)-phase diagram (Fig. 1) one recognizes the hatched area.

The origin of this hatched area is the critical point *cp*. At any temperature or pressure above *cp*, only one phase exists – the supercritical phase. At any pressure below the critical pressure (p_c) and at any temperature below the critical temperature (T_c) the phase is only subcritical. However, as will be pointed out later (Schneider) there is no boundary where the physical properties change. It is possible to start SFC or SFE in the vicinity of *cp* at or below the subcritical condition and move smoothly into the supercritical region of the phase diagram. This allows us, for example, to start below p_c but above T_c with chromatography. This would be called high pressure gas chromatography (HPGC).

Carbon dioxide is most often used for SF work but it does not dissolve polar compounds. To achieve solubility for polar compounds it is necessary to add a cosolvent or modifier, which should be completely miscible in CO_2 and which is a liquid at ambient temperatures. Methanol or other alcohols, cyclic ethers, dichloro- and trichloromethane, water and formic acid have been used. Only the last two are compatible for flame ionization detection (see below).

When utilizing modifiers, binary phase diagrams must be looked at. Actually single component pT phase diagrams can be considered irrelevant for SFC or SFE purposes, because any analyte or solvent present generates a binary phase system. When injecting a sample dissolved in a liquid, ternary or even higher phase diagrams must be considered. Fortunately in many cases we can look at an ideal diluted solution, without pondering over state parameters too much. So it is appropriate here to look at the properties of binary mixtures which can differ profoundly from single components. For

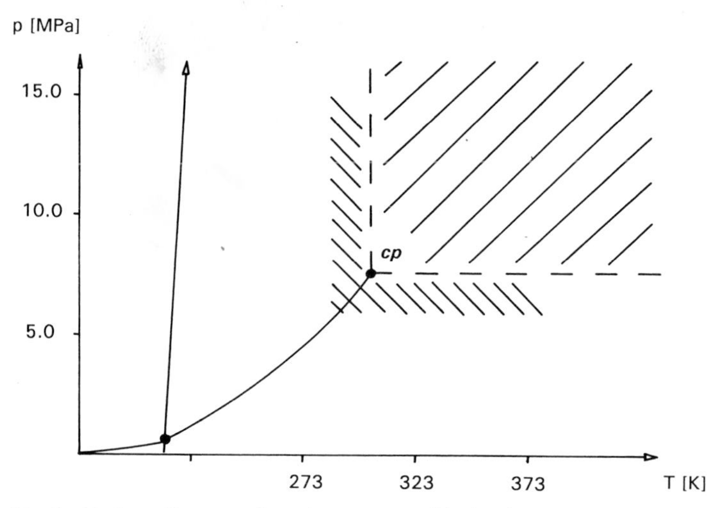

Fig. 1. pT phase diagram of a substance. *cp* critical point

example it is not possible to interpolate *cp* for CO_2/methanol mixtures from the *cp* of each component. Depending on the mole fraction (X) and temperature, there exists a distinct pressure maximum, while the critical temperatures lie between those of the pure compounds. This is very similar to water/methanol mixtures in HPLC, where the column back pressure also depends on the mole fraction and temperature and where the p/X curve has a pressure maximum.

In order to estimate if a gas or a low boiling liquid is suitable as a supercritical fluid Guldberg's rule

$$T_b = 2/3\,T_c \quad \text{or} \quad T_b \cdot 1{,}5 = T_c\,[\mathrm{K}]$$

can be applied.

It is advantageous to use supercritical fluids, the critical temperature of which is low. This favors work with analytes which decompose or react at higher temperatures.

It is common to use fluids which are gases at room temperature. A catalog of the most commonly used fluids is given in Table 1.

Supercritical fluids posses properties which resemble gases and liquids. The diffusion coefficient of a SF is somewhere in the middle between those for gases and liquids, the viscosity is similar to that of gases while the density is close to that of a liquid (Table 2).

The density changes with a maximum rate near the critical point. If only density were required one could operate near *cp*. In addition solubility depends on density. Solubility increases with pressure and small pressure differences have a marked effect on density close to the critical point. While raising pressure increases the density, raising temperature will usually decrease the density.

Table 1. Physical parameters of commonly used supercritical fluids*

	T_c[°C]	T_c[K]	T_b[K]	P_c[atm]	ϱ_c[g cm^{-3}]
CO_2	31.3	304.5	194.7	72.9	0.47
N_2O	36.5	309.7	184.7	72.5	0.45
NH_3	132.5	405.7	239.8	112.5	0.24
Ethane	32.2	305.4	184.6	48.2	0.20
Propane	96.8	369.5	226.1	42.4	0.22
Butane	152.0	425.2	272.6	37.5	0.23
CCl_2F_2	111.5	384.7	243.4	40.2	0.56
Freon 22 $CHClF_2$	96.1	369.3	232.4	49.1	0.52
H_2O	374.2	643.3	373.2	218.3	

* Data taken from [10].

Table 2. D_{12}, ϱ, and η of gases, liquids, and fluids*

	D_{12}[cm^{2}s^{-1}]	ϱ[g cm^{-3}]	η[g cm^{-1}s^{-1}]
gas	10^{-1}	10^{-3}	10^{-4}
liquid	10^{-6}	1	10^{-2}
supercritical fluid	10^{-3}	0.2 – 0.8	10^{-4}

* most data taken from [11].

Although the solubility depends on density and density changes are most profound near c_p it is not common to carry out chromatography or extract at temperatures close to *cp*. Whereas at higher temperatures the density decreases, vapor pressure and solubility increase. These combined effects produce better results for both SF techniques.

As density and thus solvent strength is very important for SF work and because the ideal gas law is not valid at high pressures other relationships which correlate pressure and density must be considered (see King/France in this book). A brief review of these connections is also given in [12]. Modern instruments provide software which uses algorithms to correlate pressure and density and there is even software for a PC available [12].

The Hildebrand solubility parameter is a means of comparing the eluting power of supercritical fluids approximately [13].

Based on theoretical considerations [14], capillary SFC requires for the same efficiency column diameters approximately one order of magnitude smaller than conventional GC and column diameters one order of magnitude higher than capillary HPLC does. This symbolizes somehow the role of SFC as a link between GC and HPLC.

Packed column SFC could take advantage of the much lower differences of inlet and outlet pressure of the column as compared to HPLC due to the low viscosities of the supercritical fluid. This is easy to demonstrate while switching from a MeOH/dichloromethane mobile phase to a MeOH/CO_2

phase. But due to the pressure drop along the column while operating under the SF mode, column efficiency also changes along the column. Packed columns for analytical SFC can be as large as a standard type HPLC column (250×4 mm) and are then most often used with a UV detector. A flame ionization detector cannot handle very high flow rates of CO_2 and therefore the column outflow must be split, or packed columns have to be decreased in size to about 100×1 mm packed stainless steel (HPLC micro columns) or even smaller packed fused silica capillary columns.

Decreasing the size of columns requires theoretical and practical considerations and a more careful set up of the instrument. A minor loss of efficiency due to instrument limitations can be accepted. As pointed out for micro LC [15] injection, connection, detector volumes can erase much of the column efficiency gained through SFC optimized columns.

Instrumentation

A basic set up for SFC can be made from either HPLC or GC instrumentation. For those who want to carryout experiments for the first time, here are some hints:

The simplest source of a supercritical fluid with ample pressure is a heated reservoir or tank filled with the desired gas. Immersing the tank in a water bath as shown in Fig. 2 can deliver pressures as high as 40 MPa. One should be careful, however, and watch the maximum pressure rating and always use a safety valve. If a thermostated water bath is available and the tank is sufficiently large the consumption of e.g. CO_2 is low, this is a "pump" with no pulsation.

Commercial instruments have syringe pumps or piston pumps. The syringe pump requires a down time for a refill after the reservoir has been emptied, but it is much easier to control and delivers high and, what is more important, low flow rates with almost no pulsation. No valves are required for the pump mode. Accordingly this pump is preferred for SFC and small volume SFE. A schematic is shown in Fig. 3.

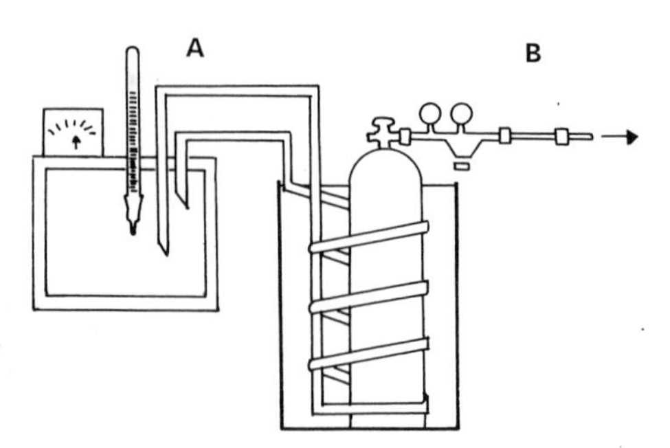

Fig. 2. Heated tank for SF delivery. A: Temperature controlled water bath. B: Gas tank in water bath

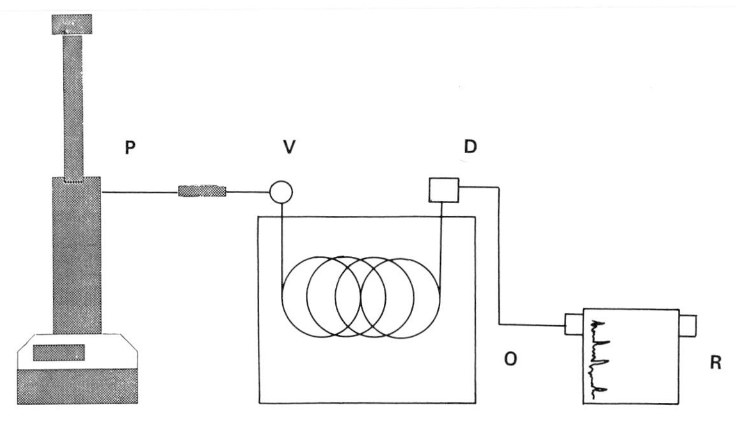

Fig. 3. SFC with syringe pump. *P*: Pump; *V*: Injection valve; *D*: Detector; *O*: Oven with column; *R*: Recorder

The piston pump delivers the fluid continuously and with electronic control, camshaft profile and pulse damper there remains almost no pulsation. Leaks from the in- and outlet valves can cause some trouble especially at very low flow rates. This pump is preferred for large volume SFE or parallel SFE of many samples.

Injection of samples into a SFC column requires a high pressure injection valve. Primarily with small inner diameter capillary columns the sample (solvent) load can be too high. Some topics pertaining to injection are addressed in a separate chapter of this book.

Developing new stationary phases for SFC either for capillary or packed columns seems to be a perpetual process. Much research concentrates as in HPLC or GC on surface coverage of the stationary phase or on selective phases e.g. enantiomer separation. Information on stationary phases can be found in the appropriate chapters of this book.

Detectors from either HPLC or GC can be used for SFC particularly with CO_2 as the mobile phase. The flame ionization detector (FID) has received the most interest for capillary and packed micro columns [16]. It is reliable, sensitive and represents the group of ionization detectors. One major disadvantage is that it cannot be used with many of the common modifiers.

The UV detector is almost universal for SF work as there are virtually no limitations in the use of the most common fluids. The UV detector is representative for the group of optical detectors. It is suited for modified mobile phases but needs a chromophore in the analyte. It is well suited for tandem detection e.g. UV-FID or UV-mass spectrometry. Scanning or photo diode array detectors provide more information than fixed wavelength UV or FID detection. Even more molecular information can be acquired through nuclear magnetic resonance or mass spectrometry detectors or SJS (supersonic jet spectroscopy). In addition to the information provided in

Table 3. Detectors for SFC

FID flame ionization
UV detector
FTIR Fourier transform infrared
MS mass spectrometer
fluorescence detector
TID thermionic detector (NPD)
FPD flame photometric detector
SJS supersonic jet spectroscopy
CD chemiluminescence detector
ECD electron capture detector
AES atomic emission spectroscopy
light scattering detector
ion mobility detector

the appropriate chapters, the recommended literature at the end of this book might also be of interest.

Restrictors

In order to maintain sufficiently high pressures either with SFC or SFE the flow through the apparatus must be limited at the outlet with a restrictor. Crimped stainless steel or platinum capillaries were used at first. Good heat transfer and stability are the advantages of this type; however very often the flow is not directed and therefore not easy to adjust. In addition, they are not chemically inert for certain compounds, especially when the restrictor is heated.

Fused silica capillary restrictors became more common and a small selection is shown in Fig. 4. Type A is just a piece of fused silica with an inner diameter of 5 – 15 μm for SFC and 15 – 50 μm for SFE. The tapered restrictor (B) is easily made but not very rigid. The frit restrictor (C) has the advantage of "many holes" and does not plug as easily as a one hole restrictor does. In the author's laboratory, the Guthrie restrictor is the most often used for SFC work (Type D) [17]. With some experience it can be made in less than an hour and with a desired flow rate.

All of these restrictors have one major shortcoming: They change the flow dependent on the pressure of the supercritical fluid. This is especially bad for chromatography purposes due to the influence of the mobile phase velocity on column efficiency. Only larger columns (standard HPLC size) or SFE instruments with high SF flow rates can be equipped with valves to control the flow [18].

Due to the Joule-Thomson cooling effect, restrictors can be obstructed by solvent freezing or extract deposition within the restrictor. Heating of the

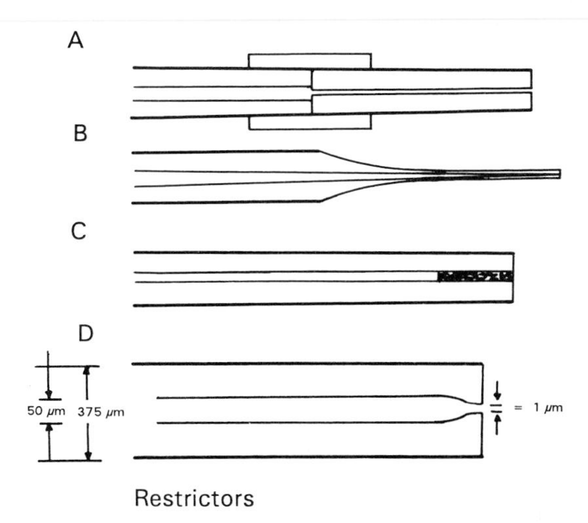

Fig. 4. Common restrictors for SFE and SFC. (For details see text)

restrictor is therefore necessary. Sometimes the collection solvent provides sufficient heat if extraction periods are not too long and the flow is not too high.

A hair dryer, or a heat gun, or even better, a thermostated heat box can be used to avoid the freezing problem.

A basic SFE setup for modified supercritical carbon dioxide extraction is shown in Fig. 5. Three basic components are needed for SFE: The fluid delivering system, the extraction cell in a heater (oven or heating block) and a restrictor.

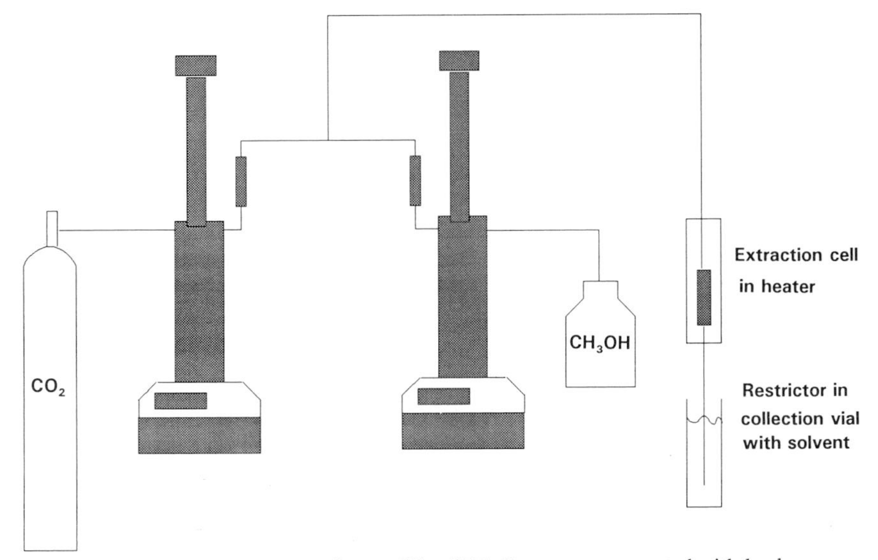

Fig. 5. Dual syringe pump system for modifier SFE. Pumps are connected with backpressure control valves and mixing tee

A very simple SFE apparatus has been shown by J. King. A steel container is filled with dry ice and the sample. The container can be moderately heated and this raises the pressure sufficiently above c_p. Certainly pressure depends then on filling and temperature, but it is well suited for screening purposes.

SFE apparatus can be connected on- and offline to other instruments most commonly those used for chromatography. Details about these techniques can be found in the special chapters in this book.

References

1. Klesper E, Corwin AH, Turner DA (1962) J Org Chem 27:700
2. Giddings JC, Myers M, McLaren L, Keller R (1968) Science 162:67
3. Sie ST, Rijnders GWA (1967) Anal Chim Acta 38:31
4. Gouw TH, Jentoft RE (1972) J Chromatogr 68:303
5. Novotny M, Springston SR, Peaden PA, Fjeldsted JC, Lee ML (1981) Anal Chem 53:407 A
6. Hawthorne StB (1990) Anal Chem 62:633 A
7. McHugh MA, Krukonis VJ (1986) Supercritical Fluid Extraction: Principles and Practice. Butterworths, Boston
8. Johnston KP (1989) Penninger JML (eds) Supercritical Fluid Science and Technology. ACS Symposium Series 406, Washington
9. Stahl E, Quirin KW, Gerard D (1987) Verdichtete Gase zur Extraktion und Raffination. Springer, Berlin Heidelberg New York
10. Weast RC (Ed) CRC Handbook of Chemistry and Physics (1974) 55th Ed. Chemical Rubber Company, Cleveland, OH
11. Schneider GM (1978) Angew Chem Int Ed 17::716
12. SF-Solver, 1-2 – 1-7, ISCO. Lincoln Nebraska, 4/91
13. Giddings JC, Myers M, McLaren L, Keller R (1986) Science 162:67
14. Fields SM, Kong RC, Fjeldsted JC, Lee ML, Peaden PA (1984) JHRC & CC 7:312
15. Wenclawiak BW (1984) Microchimica acta
16. Richter BE, Bornhop DJ, Swanson JT, Wangsgaard JG, Andersen MR (1989) J Chromatogr, Sci 27:303
17. Guthrie EJ, Schwartz HE (1986) J Chromatogr Sci 24:236
18. Jahn KR, Wenclawiak BW (1987) Anal Chem 59:382

2 Physico-chemical Principles of Supercritical Fluid Separation Processes

GERHARD M. SCHNEIDER

2.1 Introduction

Fluid solvents predominantly in the critical and supercritical ranges are of considerable interest for many fields e.g. for some new separation methods such as Supercritical Fluid Extraction (SFE) and Supercritical Fluid Chromatography (SFC); some selected monographs and reviews and many references are e.g. given in Refs. [14–17].

Fundamental investigations in these fields have also been a focal point of the scientific activities of the author's laboratory at the University of Bochum for a long time. Several reviews [1–16] as well as many dissertations and original papers have already appeared. The present article is a summary and an extension of similar preceding papers [15, 16] where the material has been discussed in more detail and numerous references have been given. In the following some basic aspects will be reviewed that are of interest for the understanding of the physicochemical principles of SFE and SFC processes.

For every extraction process (including SFE) the detailed knowledge of many thermodynamic and transport properties of the solvent phase, the solutes and all mixtures involved is of primary importance e.g. pVTx data, phase equilibria, solubilities, viscosities, diffusion coefficients, surface tensions etc. The same holds for chromatographic applications; here, however, the accent is on very dilute solutions.

From a physico-chemical point of view the most important informations that can be obtained from a chromatographic peak are

- its *position* (e.g. characterized by the retention time t_R or the retention volume) and
- its *profile* (e.g. characterized by the variance in case of a Gaussian peak).

For the *peak position*, the following relation holds with the assumption of a partition equilibrium of a component i between the stationary (stat) and the mobile (mob) phase

$$k_i' \equiv \frac{c_i^{stat}}{c_i^{mob}} \cdot \frac{V^{stat}}{V^{mob}} = \frac{t_{Ri} - t_o}{t_o} \qquad (1)$$

where k_i' = capacity ratio of sample i; t_{Ri} = retention time of sample i; t_0 = retention time of an inert sample; c_i^{stat} and c_i^{mob} = concentration of the sample i in the stationary and mobile phase, respectively; V^{stat} and

V^{mob} = volume of the stationary and the mobile phase, respectively. Since

$$\frac{c_i^{stat}}{c_i^{mob}} \equiv K_i \tag{2}$$

where K_i is the partition coefficient, the capacity ratio k_i' and consequently the retention time t_{Ri} are determined by thermodynamics e.g. by solubilities, phase equilibrium and its dependences on temperature, pressure, added substances (moderators) etc. Relations that are equivalent to Eq. (1) and (2) hold for other retention mechanisms e.g. in adsorption chromatography. Some basic aspects of this field will be reviewed in Sect. 3.

The *peak profile*, however, is mainly determined by transport properties (predominantly diffusion coefficients). These topics will be discussed in Sects. 2 and 4.

2.2 Physico-chemical Properties of Pure Supercritical Solvents

Carbon dioxide is the supercritical solvent most frequently used at present in SFE and SFC. Since it is a perfect model substance for such a fluid, some of its relevant physico-chemical properties will be presented and discussed briefly in the following.

2.2.1 Thermodynamic Properties

In Fig. 1a the p(T) phase diagram of carbon dioxide is presented. On the vapor pressure curve (lg) a liquid (l) and a gaseous (g) phase coexist. With increasing pressure the vapor pressure curve rises and ends at the critical point CP (304.21 K, 73.825 bar, 0.466 g cm^{-3} [43]) where the two coexisting phases (l and g) become identical. With decreasing pressures it ends at the triple point Tr (216.58 K, 5.185 bar [43]) were solid, liquid and gaseous carbon dioxide are in equilibrium. The sublimation pressure curve (gs) and the melting pressure curve (ls) are not important within the scope of the present article.

In Fig. 1a the ranges of temperatures and pressures are additionally marked where separation methods (such as distillation, liquid extraction, GC, SFE, SFC) are normally performed. Here SFE and SFC have to be attributed to conditions of temperatures and pressures exceeding the critical values respectively. In principle this definition is somewhat vague since (at least in SFE) only the ranges are relevant where the mixtures involved are critical or supercritical and not the pure solvent.

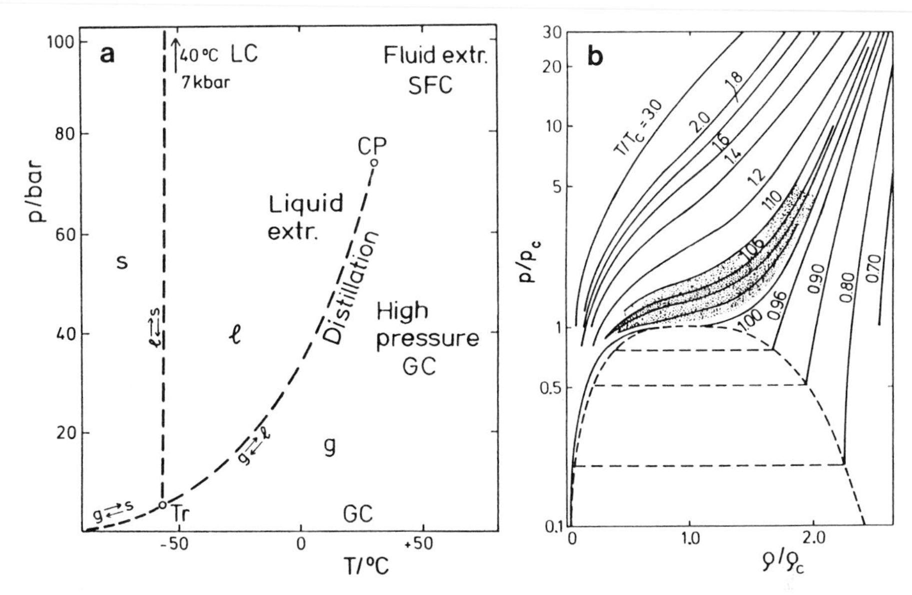

Fig. 1. **a** p(T) phase diagram and **b** p(ϱ) isotherms of pure carbon dioxide (p = pressure, T = (absolute) temperature, ϱ = density, CP = critical point, p_c = critical pressure, T_c = (absolute) critical temperature, ϱ_c = critical density, Tr = triple point (here gls), l = liquid, g = gaseous, s = solid) (according to [43])

In the reduced p(ϱ) diagram in Fig. 1 b the range relevant for SFE and SFC is dotted. Here important density variations are induced by small pressure changes at constant temperatures i.e. the compressibility β that is defined by

$$\beta \equiv -\frac{1}{V_m}\cdot\left(\frac{\partial V_m}{\partial p}\right)_T = \frac{1}{\varrho}\cdot\left(\frac{\partial \varrho}{\partial p}\right)_T \tag{3}$$

is high and at CP itself $\beta = \infty$.

2.2.2 Dielectric Properties

In Fig. 2, the static dielectric constant (relative permittivity) ε of carbon dioxide and argon respectively is plotted against pressure p up to 2 kbar. Even at the highest pressures corresponding to liquid-like densities $\varepsilon(CO_2)$ is smaller than 1.8 and thus equals about that of a liquid alkane (e.g. heptane). Since CO_2 molecules do not have any permanent electrical dipole moments, the polarisation is more or less restricted to the contributions of the electrons and the nuclei; thus typical solvation effects are not possible and the intermolecular interactions are of van der Waals type, quadrupolar interactions being negligible.

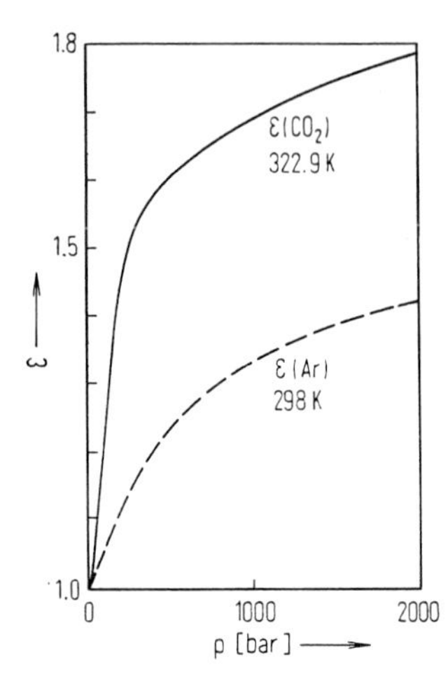

Fig. 2. (Static) dielectric constant (permittivity) ε of pure carbon dioxide [46] and argon [47] as a function of pressure (see also [5])

This is evidence that the solvent power of supercritical condensed carbon dioxide is rather poor and corresponds to that of a typical hydrocarbon only. As a consequence carbon dioxide is not a "supersolvent" at all but on the contrary a rather poor solvent and adequate for substances with low or medium molar mass and polarity (such as many natural products) only. The advantage of supercritical solvents of this type, however, consists of the possibility of varying the solvent power not only by a temperature change or by additives (as for liquid solvents) but in addition by an adequate pressure or density change very easily and over orders of magnitude.

2.2.3 Transport Properties

As can be seen from Fig. 3 supercritical solvents exhibit rather small viscosities and high self-diffusion coefficients which resemble more those

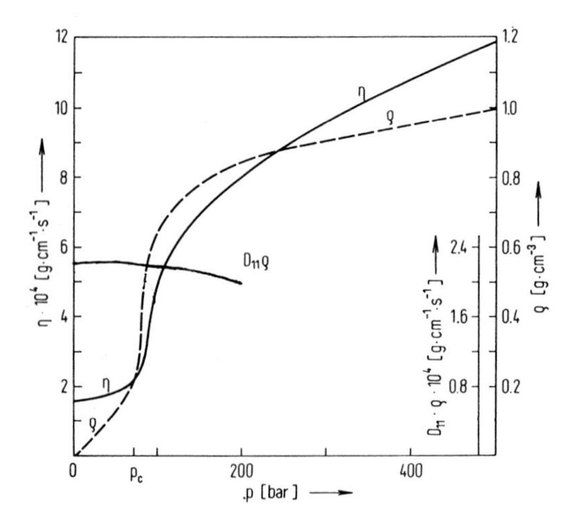

Fig. 3. Density ϱ, viscosity η and $D_{11} \cdot \varrho$ of pure CO_2 as a function of pressure at 40 °C (for $D_{11} \cdot \varrho$ at 50 °C) (according to literature data [24], see also [5, 50])

of gases than those of liquids. Similar properties are also found in dilute solutions of low-volatile samples in supercritical solvents of this kind. In addition, the supercritical solvent is also considerably soluble in the coexisting liquid phase making viscosities lower and diffusion coefficients higher. In this way, mass transfer and the equilibration rate are very much improved for SFE in comparison with liquid solvent extraction (see Sect. 4).

2.3 Phase Equilibria of Fluid Mixtures

2.3.1 Phase-Theoretical Aspects

As already shown in several preceding review papers [1–16] three different types of heterogeneous phase equilibria are found in fluid mixtures namely

- liquid-liquid equilibria (ll),
- liquid-gas equilibria (lg) and
- gas-gas equilibria (gg).

Figure 4 shows some schematic pTx diagrams (upper row) and the corresponding p(T) projections (lower row) for a binary mixture; here p = total pressure, T = temperature, x = mole fraction. Most important for the

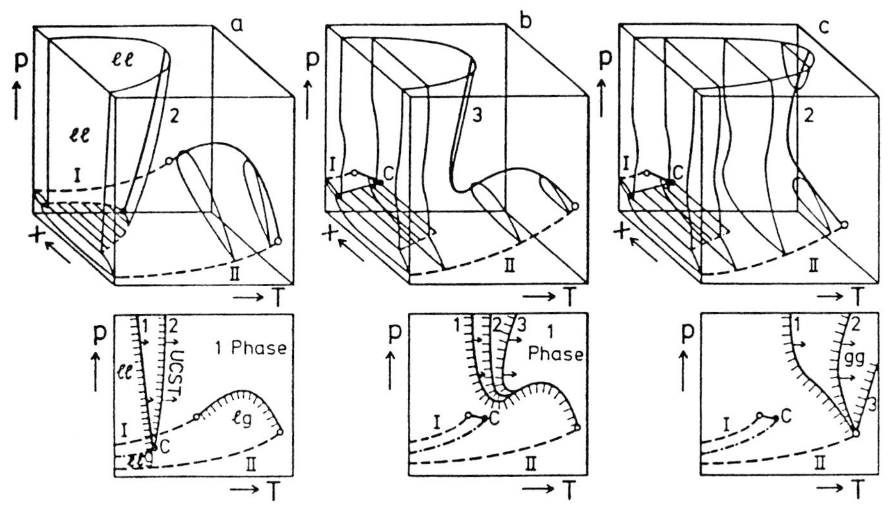

Fig. 4. pTx diagrams (upper row) and p(T) projections (lower row) for binary mixtures (schematically, see text: ——— = p(x) isotherm; ——— = critical curve of binary mixtures; – – – – = vapor pressure curve of pure components; – · – · – · = three-phase line llg; o = critical point of pure component (CP); C = critical endpoint; UCST = upper critical solution temperature; lg = liquid-gas; ll = liquid-liquid; gg = gas-gas; for details see e.g. [5, 6, 14, 16, 17])

classification of fluid phase equilibria is the shape of the different critical curves. A critical curve connects all binary critical points in the pTx space; here two coexisting fluid phases become identical (e.g. at the extreme values of the isothermal p(x) or isobaric T(x) sections; for details see e.g. [2, 5, 6, 16, 17]).

The diagram in Fig. 4a belongs to a system showing a non-interrupted critical line lg and at the same time a liquid-liquid phase separation (ll) at much lower temperatures. For the following Fig. 4b and c, the liquid-liquid miscibility gap is more and more shifted to higher temperatures: In Fig. 4b, the critical line ll merges continuously into the critical line lg; in Fig. 4c, no pressure maximum or minimum is found any longer and the systems shown have to be attributed to the so-called gas-gas equilibrium type of the second (type 2 in Fig. 4c) or first kind (type 3 in Fig. 4c), respectively. Many different transition types exist.

The critical curves shown in Fig. 4 are also of primary interest for SFE and SFC: Under the conditions of temperature and pressure on the right-hand side of the critical curves, the supercritical solvent I and the low volatile component II are miscible in all proportions whereas phase splitting exists on the left hand side which is of interest for dissolution and separation processes, respectively.

The shape of characteristic isobaric T(x) and isothermal p(x) sections can be deduced from the three-dimensional phase diagrams in the upper row of Fig. 4; for p(x) sections also see Fig. 11c (for details see e.g. [5, 6, 16]).

2.3.2 Classification of Binary Critical Phase Behavior by "Families"

The usefulness of the classification given in Fig. 4 is demonstrated on some solvent-solute "families" in Figs. 5 and 6. In such a "family" one component (e.g. the solvent I) is maintained constant whereas the component II is varied systematically (e.g. within a homologous series).

As an example some members of the CO_2+alkane family are presented in Fig. 5. The figure demonstrates that the system octane+CO_2 exhibits liquid-liquid immiscibility and belongs to type 2 in Fig. 4a, for hexadecane +CO_2 and squalane+CO_2 continuous transitions between lg and ll critical states on the critical line are found according to type 3 in Fig. 4b; as has been shown elsewhere the system tridecane+CO_2 corresponds to the transition type between both.

For the CHF_3+alkane family [36] this transition happens between hexane+CHF_3 and octane+CHF_3 (see Fig. 6) whereas Ar+CHF_3 and N_2+CHF_3 correspond to type 1 given schematically in Fig. 4c. Systematic changes of the shapes of the critical curves are also found for other families where a constant component I e.g. methane, ethane, ethylene, nitrogen, water, tetrafluoromethane, helium, neon, argon, sulfurhexafluoride, metha-

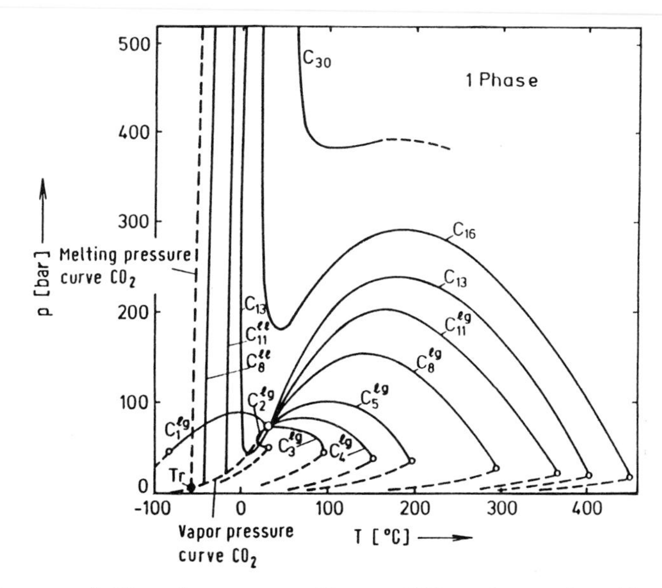

Fig. 5. p(T) phase diagrams of CO_2+alkane systems ($C_n = n-C_2H_{2n+2}$ for n = 1 to 16; C_{30} = 2,6,10,15,19,23-hexamethyl-tetracosane (squalane); for details and references see [5])

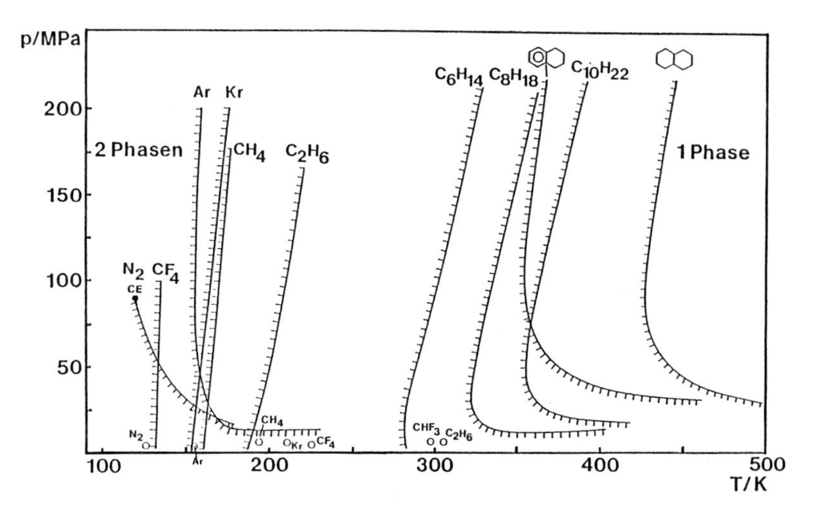

Fig. 6. p(T) projections of the critical curves of some binary trifluoromethane systems (for symbols see legend of Fig. 4) (according to Kulka [36])

nol, ammonia etc. is combined with a second component II that is systematically varied e.g. within a homologous series; for reviews and references see [1–17]. For a compilation of the literature concerning experimental methods and systems investigated (1978–1987) see [48] and of gas-liquid critical data for mixtures see [49].

2.3.3 Ternary Systems

Measurements on ternary systems are instructive in order to study

- the influence of a mixed solvent on the SFE extraction of one low-volatile substance or
- the extractive effect of a supercritical solvent (here CO_2) on the separation of two low-volatile substances.

Examples for the *first* effect are shown in Fig. 7a and b [45]. The ternary isothermal p(w) Gibbs prism in Fig. 7a demonstrates that $CClF_3$ added to

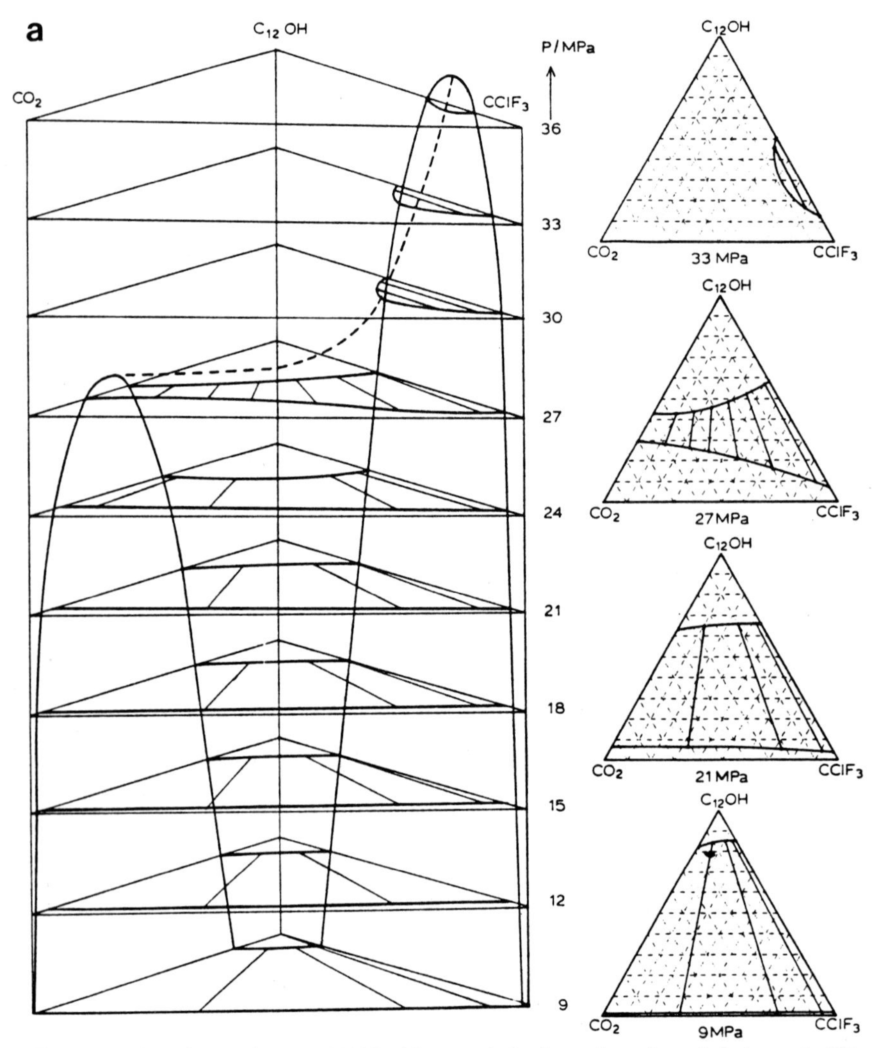

Fig. 7a, b. p(w) phase prisms at 392 K with some isobaric sections for **a** 1-dodecanol + CO_2 + $CClF_3$ and **b** 1-dodeccanol + CO_2 + N_2 (according to Katzenski et al. [46])

the supercritical solvent CO_2 decreases the solvent power of the supercritical solvent phase at rather high mass fractions only. For an addition of N_2 (see Fig. 7b) the decrease of solvent power, however, is important even at the lowest concentrations; N_2 is therefore sometimes used in extractions with supercritical CO_2 as a separating agent. For both, $CClF_3$ and N_2, as supercritical solvents addition of CO_2 increases the solvent power; thus CO_2 improves the mutual miscibility and acts as an entrainer (enhancer) in these cases.

The *second* effect is demonstrated in Fig. 8a and b. Whereas the solvent power of CO_2 for a mixture of 1-dodecanol+1-hexadecanol decreases monotonically by adding 1-hexadecanol to the 1-dodecanol (Fig. 8b), the

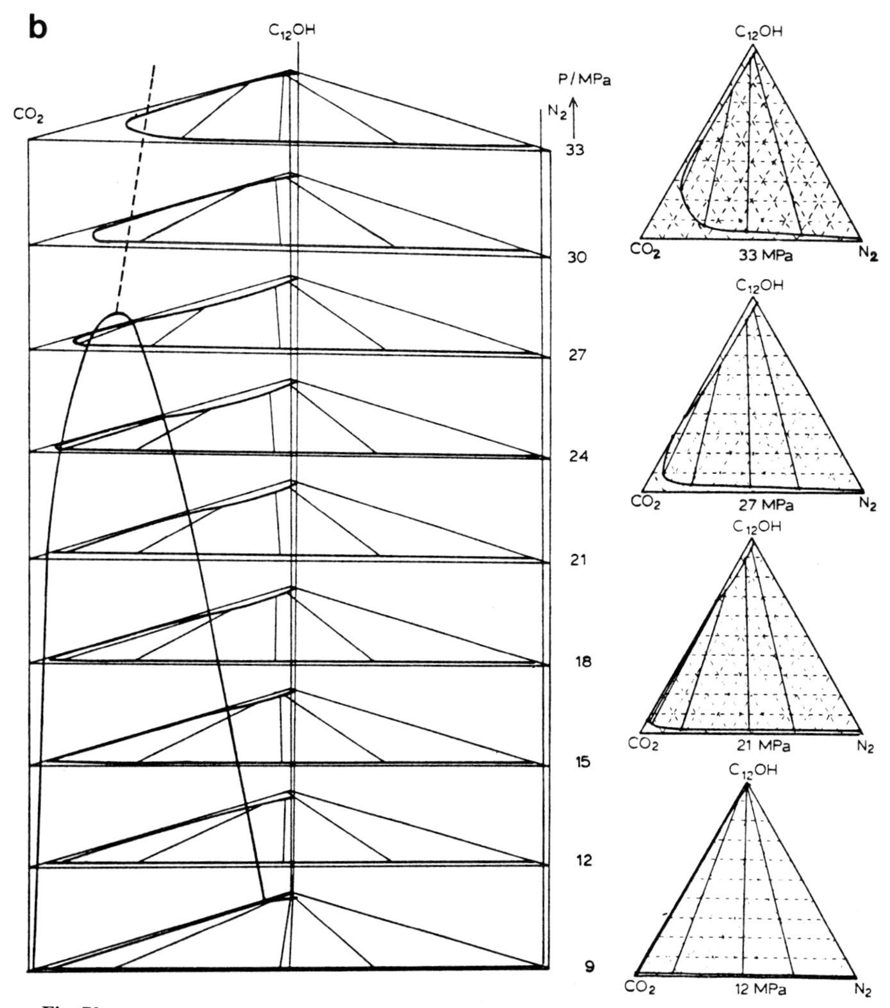

Fig. 7b

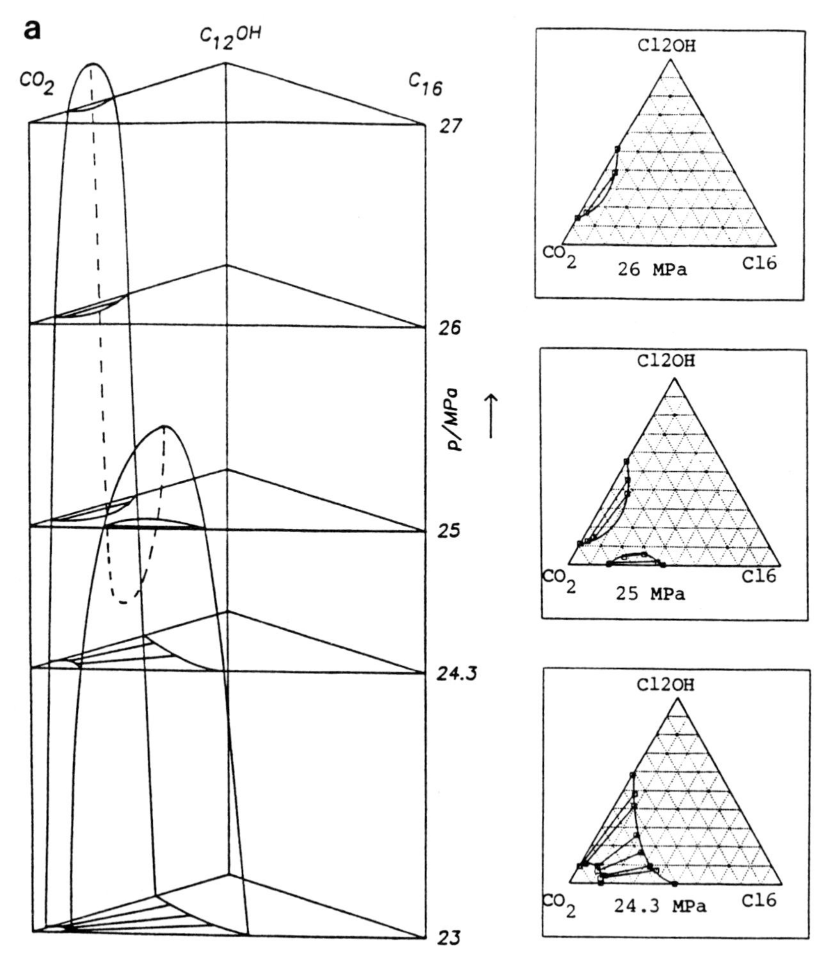

Fig. 8 a, b. p(w) phase prisms at 393.2 K with some isobaric sections for **a** 1-dodecanol + hexadecane + CO_2 and **b** 1-dodecanol + 1-hexadecanol + CO_2 (according to Hölscher et al. [38])

ternary critical curve of the system CO_2 + 1-dodecanol + hexadecane (Fig. 8 a) runs through a distinct pressure minimum giving evidence for a so-called cosolvency effect where a mixture of two components (here 1-dodecanol and hexadecane) has a higher solubility in a solvent (here CO_2) than each of the two pure components alone. Cosolvency phenomena are found in polymer solutions and are mainly caused by entropy effects; for details see [37, 38].

Both effects are of interest in SFE e.g. for the design of an extraction process and the construction of an extraction plant as well as in SFC e.g. for the use of mixed mobile phases and gradient techniques. The above-mentioned findings demonstrate that these effects are often not linear.

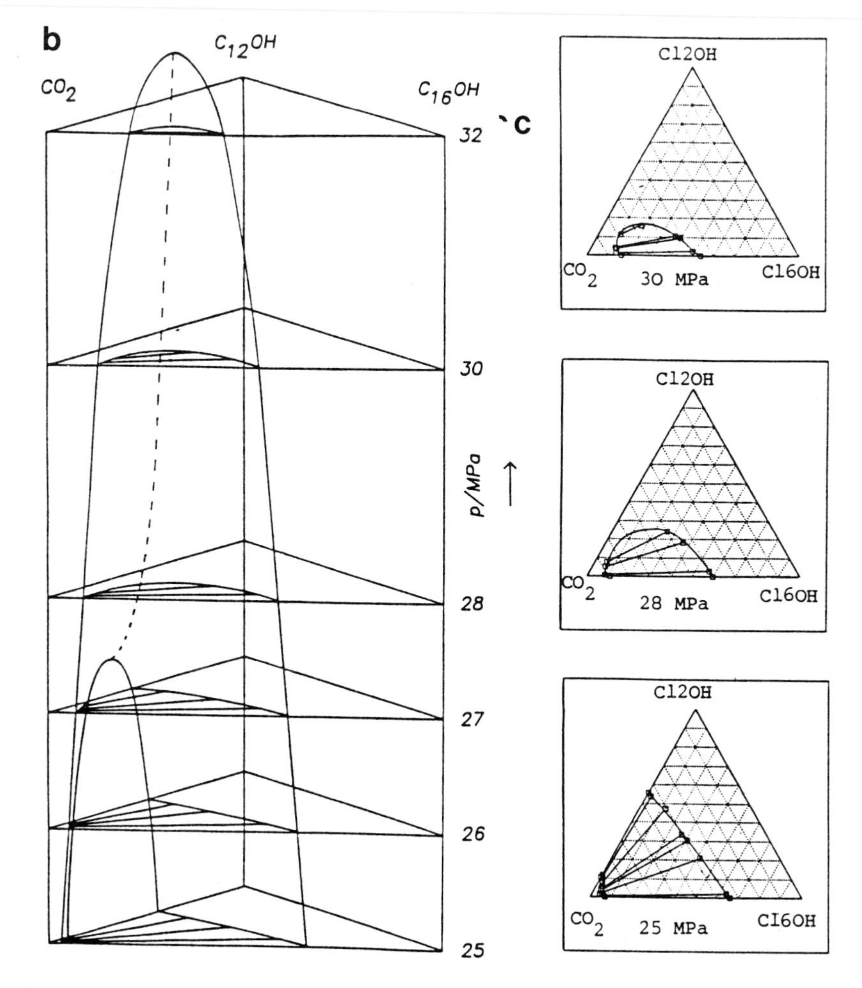

Fig. 8b

2.3.4 Quaternary Systems

As a rule selectivity is poor for extractions with supercritical CO_2 and other supercritical solvents which makes it necessary to use so-called moderators. Their action can be studied in a quaternary system consisting of the supercritical solvent (e.g. CO_2), two low-volatile substances to be separated (e.g. 1-dodecanol and hexadecane) and a low-volatile moderator (e.g. 1,8-octanediol or dotriacontane).

For both cases the so-called separation factor

$$\alpha = K_{\text{hexadecane}} / K_{\text{dodecanol}} \tag{4}$$

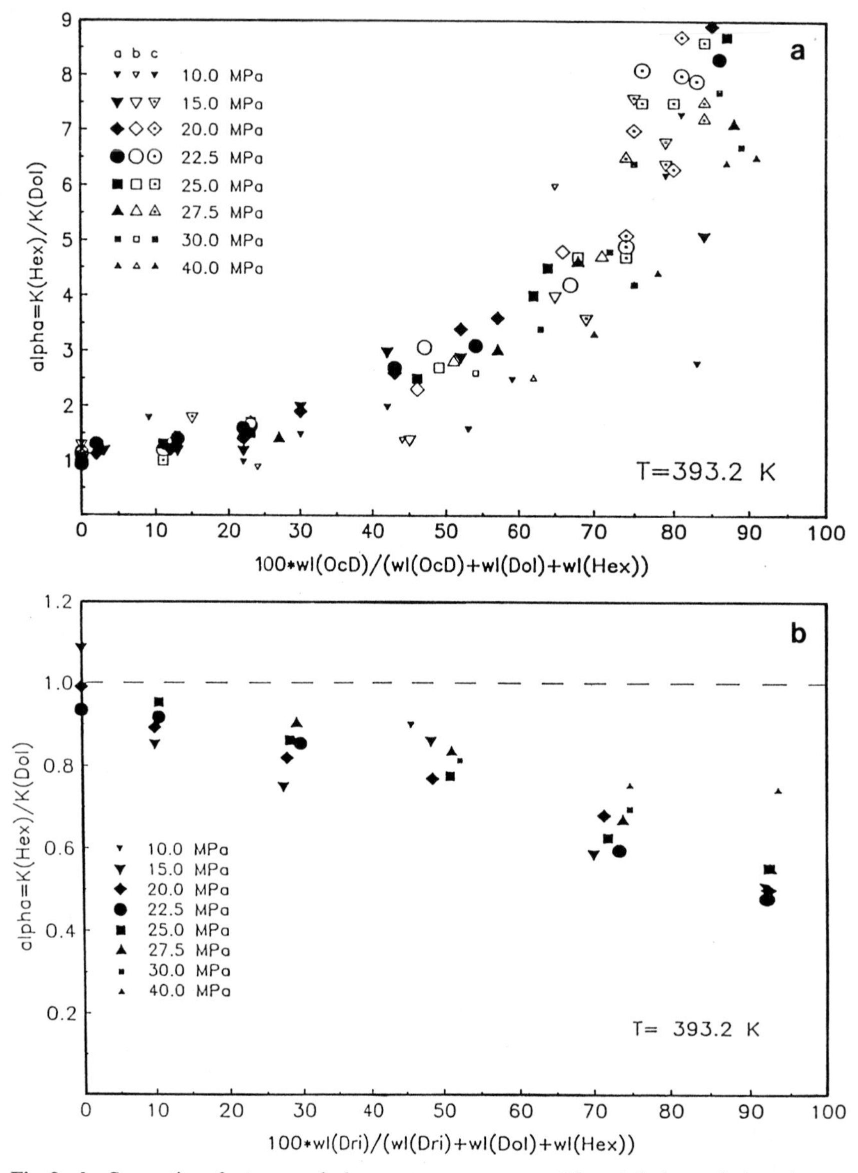

Fig. 9a, b. Separation factor α of the quaternary system CO_2 + 1-dodecanol + hexadecane + moderator at 393.2 K as a function of the mass fraction w of the moderator (CO_2 free basis). **a** moderator = 1,8-octanediol (w(dodecanol)/w(hexadecane) = 1 : 1 (a), 2 : 1 (b) and 1 : 3 (c); **b** moderator = dotriacontane (w(dodecanol)/w(hexadecane) = 1 : 1) (according to Spee [37])

is plotted against the relative mass fraction of the moderator on a solvent-free basis in Fig. 9a and b. Whereas α typically equals about unity for the ternary system without addition of moderator, values up to about 9 and down to about 0.5 are found with increasing mass fractions of 1,8-octanediol (see Fig. 9a) and dotriacontane (see Fig. 9b), respectively, giving evidence of a distinct extractive effect of the added moderator. The effects can be explained qualitatively by the fact that the (polar) 1,8-octanediol preferentially holds the (polar) component 1-dodecanol back in the CO_2- poor phase whereas the (unpolar) dotriacontane does the same with the (unpolar) component hexadecane. For a detailed discussion of this example see [37].

The change of selectivity by the use of mixed solvents as mobile phases or by the addition of adequate moderators will be of considerable importance for the future development and application of SFE and SFC.

2.3.5 Solubility of Solids in Supercritical Solvents

In applied SFE, solutes are quite often solids (e.g. many natural products). In these cases the phase diagrams can be very difficult since crystallization and fluid phase equilibria overlap.

The two most important cases are shown in Fig. 10. In Fig. 10a, the three-phase line s_2gl, where pure solid component 2 is in equilibrium with a saturated liquid and a coexisting vapor phase, is situated at pressures below the critical curve lg of the binary system; a characteristic isothermal p(x) diagram for $T_1 > T_{c1}$ is shown in Fig. 10c, case a. For the system in Fig. 10b, however, the three-phase line is shifted to higher pressures and intersects the critical curve lg twice at the critical endpoints A and B. In Fig. 10c, typical isothermal p(x) sections for $T_1 = \text{const}$ (type a) and $T_2 = \text{const}$ (type b) are represented; here the curve d corresponds to the solubility of pure solid 2 in the supercritical solvent 1 (e.g. CO_2) for $T_A < T < T_B$.

Of special interest are the isothermal p(x) diagrams that correspond exactly to sections through the critical endpoints A and B. Here the fluid-fluid miscibility gap with the critical point CP vanishes resulting in an horizontal tangent to the p(x) isotherm with $(\partial x_2/\partial p)_T = \infty$; this corresponds to a remarkable pressure effect. For details and more complex phase diagrams see [44].

2.3.6 Isothermal Pressure (Concentration) Phase Diagrams

In all the phase diagrams presented, mole fractions x were used to characterize composition. In separation processes, however, concentrations in mol/volume (e.g. $mol \cdot m^{-3}$) or in mass/volume (e.g. $kg \cdot m^{-3}$) are more important since partition coefficients, capacity ratios etc. are defined using concentrations instead of mole fractions (see Eqs. (1) and (2)).

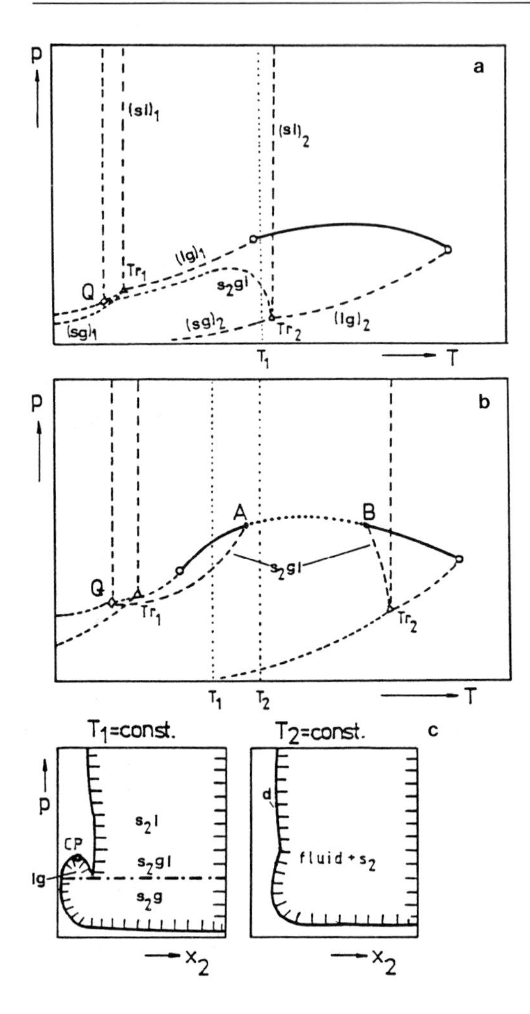

Fig. 10a–c. p(T) diagrams (**a, b**) and selected p(x) sections (**c**) for binary systems with occurrence of a solid phase in the supercritical region fluid-fluid in simple cases (schematically, see text; for symbols see legend of Fig. 4) (according to [5])

The differences between the isothermal p(x) and p(c) phase diagrams are very important especially at low pressures. This is demonstrated in Fig. 11. Figure 11a and c correspond to the type of phase behavior such as shown in Fig. 4a. Point B represents pure liquid component 2 and A the coexisting vapor phase. Both the mole fractions of component 2 equal unity but the corresponding concentrations (e.g. in mol/volume) are different in most cases even by orders of magnitude. This is shown in Fig. 11e where the total pressure p is plotted against log c_2. By increasing the pressure (e.g. by adding the supercritical solvent 1) c_2^g normally increases by many orders of magnitude; this increase is higher the lower the vapor pressure of pure component 2 is. For pressures above that of the binary critical point (D) component 2 is completely soluble in the supercritical solvent 1. This gives evidence for the often enormous change of solvent power of supercritical phases with increasing pressures and densities in SFE and SFC.

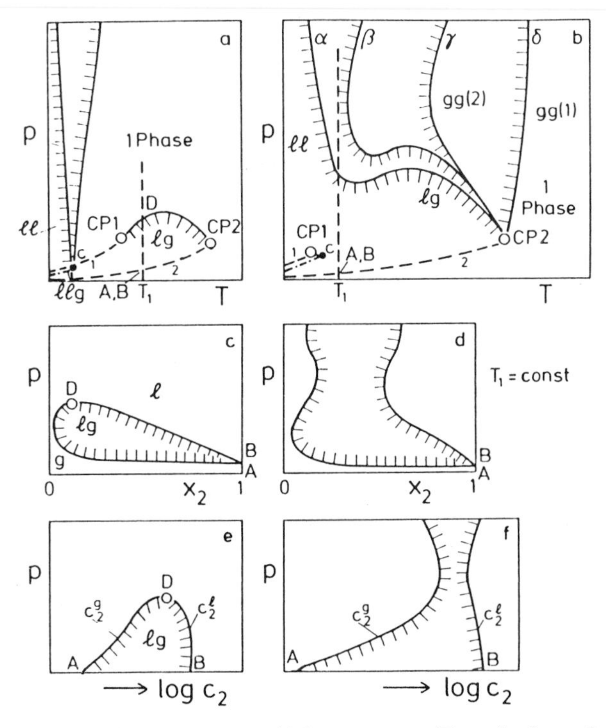

Fig. 11 a–f. Phase behaviour of fluid binary systems at high pressures: p(T) projections of phase diagrams **(a, b)** and selected isothermal p(x) **(c, d)** and p(log c) sections **(e, f)** (schematically; for symbols see legend of Fig. 4; for A, B see text) (see e.g. [7])

Figure 11 d and f show the $p(x_2)$ and $p(\log c_2)$ phase diagrams that correspond to an isothermal section at $T = T_1 = \text{const}$ for type β in Fig. 11 b (the same type is also presented in Fig. 4 b). As does Fig. 11 e, Fig. 11 f demonstrates the considerable increase of c_2^g with increasing pressure p of the supercritical solvent phase.

Pressure (concentration) diagrams can in principle be calculated from pressure (mole fraction) diagrams with the knowledge of an equation of state (EOS) for the pure supercritical solvent (at very low concentrations of component 2) or of the mixtures (at high concentrations of component 2; Sect. 3.7).

In some cases, the concentrations can be determined directly e.g. from spectroscopic measurements using Lambert-Beer's law. The $p(\log c_2)$ isotherm of the system $CO_2(1)$+3-hexanol(2) at 323.4 K in Fig. 12 has been determined from near-infrared spectroscopic measurements. Since here the separate determination of the concentrations of monomeric (mon) and associated (ass) 3-hexanol species in the fluid solution has been possible, the $p(\log c_{2\,mon})$ and the $p(\log c_{2\,ass})$ curves are also given in Fig. 12; for details see [39].

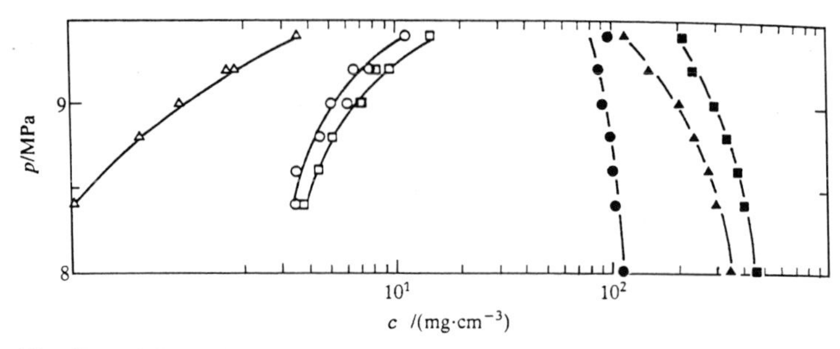

Fig. 12. p($_{10}$log c) isotherms for 3-hexanol in supercritical carbon dioxide at 323.4 K, ○ = monomeric, g; ● = monomeric 1; △ = associated, g; ▲ = associated, l; □ = total, g; ■ = total, l) according to Friedrich et al. [39])

2.3.7 Calculation and Correlation of Fluid Phase Equilibria

The aim of activities in this field is the numerical determination of phase equilibrium data of mixtures including critical phenomena from properties of the pure components. Here the use of equations of state (EOS) is at present the most promising approach whereas empirical correlations, corresponding states approaches, lattice theories etc. are currently much less important. The EOS chosen must describe the pVT behavior of the pure components and the mixtures quantitatively sufficiently well (if possible in both coexisting phases), contain only few parameters easily obtainable from experiments and allow the calculation of the phase equilibria and critical phenomena at elevated pressures at least semiquantitatively.

The EOS most widely used at present is the Redlich-Kwong equation of state (R. K.) possibly with some modifications e.g. the Carnahan-Starling-Redlich-Kwong EOS (C. S. R. K.). The R. K. equation of state is given by

$$p = \frac{RT}{V_m - b} - \frac{a}{\sqrt{T}\, V_m (V_m + b)} \tag{5}$$

Here a and b can be calculated from the critical values of temperature and pressure.

For mixtures the following mixing rules are normally used

$$a = a_{11} x_1^2 + 2 a_{12} x_1 x_2 + a_{22} x_2^2 \tag{6a}$$

$$b = b_{11} x_1^2 + 2 b_{12} x_1 x_2 + b_{22} x_2^2 \tag{6b}$$

$$a_{12} = \theta \sqrt{a_{11} a_{22}} \tag{6c}$$

$$b_{12} = \zeta \cdot \frac{b_{11} + b_{22}}{2} \tag{6d}$$

Here θ and ζ characterize the deviations of a from the geometric mean and of b from the arithmetic mean of the pure components respectively. Many other evenmore sophisticated equations of state (e.g. the Deiters equation) and mixing rules exist that give better agreement with experiments.

For the molar Gibbs energy G_m of a binary mixture of given mole fraction x_1 at temperature T and pressure p the following relation holds

$$G_m(p,T) = x_1 \cdot G^*_{m1}(p^+,T) + x_2 \cdot G^*_{m2}(p^+,T) + R \cdot T \cdot (x_1 \cdot \ln x_1 + x_2 \cdot \ln x_2) - \int_{V_m^+}^{V_m} p(T, V_m, x_1)\, dV_m + p \cdot V_m - R \cdot T \quad (7)$$

Here p^+ is a rather low pressure and V_m^+ the corresponding molar volume where ideal gas behavior can be assumed. The first and the second terms in Eq. (7) give the contributions of the pure components, the third term the entropy of mixing for ideal gas behavior and the fourth term the change of G_m on compression of the mixture to the pressure p or the corresponding molar volume V_m respectively; in this fourth term the EOS is used. The last two terms result from replacing A_m by G_m in the derivation of Eq. (7).

The chemical potential μ_i of component i that is defined by

$$\mu_i = \left(\frac{\partial G}{\partial n_i}\right)_{T,p,n_{j(j \neq i)}} \quad (8)$$

is then calculated from Eqs. (7) and (8) or by using

$$\mu_i = G_m + (1 - x_i) \cdot \left(\frac{\partial G_m}{\partial x_i}\right)_{p,T} \quad (9)$$

Finally the mole fractions x_i' and x_i'' of two coexisting phases ′ and ″ are obtained from

$$\mu_i' = \mu_i'' \qquad \text{for } p, T = \text{const} \quad (10)$$

or an equivalent relation.

If a supercritical fluid phase coexists with a solid, the chemical potential of the solid component (pure or in a solid mixture) is used in Eq. (10) [41, 42].

An interesting example for such a calculation is shown in Fig. 13 where the concentrations of saturated solutions of solid adamantane in supercritical CO_2 is given as a function of pressure [13, 41, 42]. The diagram demonstrates that the R.K. equation of state is a good approximation up to about 30 MPa which is more or less an upper limit for the pressure range in SFE and SFC. For higher pressures, however, a more sophisticated EOS must be used (e.g. that of Deiters).

For an advanced treatment of the calculation and correlation of fluid phase equilibria including many details and examples see e.g. [40]. It should

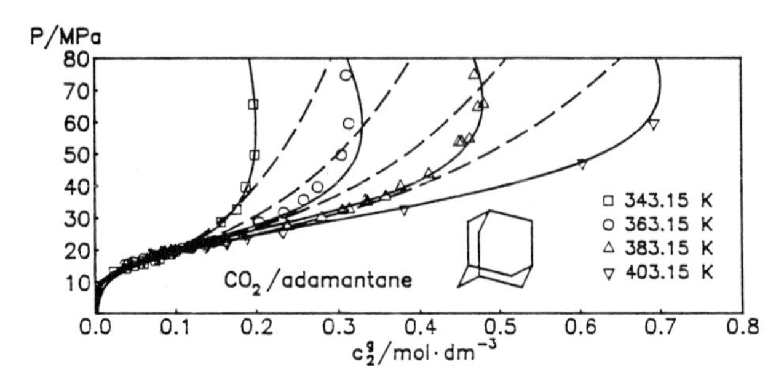

Fig. 13. Isothermal $p(c_2)$ curves for the solubility of solid adamantane in supercritical CO_2 (data points = experimental; – – – – = calculated with Redlich-Kwong equation of state, ——— = calculated with non-cubic equation of state; (according to Deiters et al. [13, 41])

be mentioned here that the R. K. equation of state and most others used are analytical. Since normally no effects extremely near to critical states have to be considered it is not necessary to use non-analytical relations that would be in accordance with the accurate critical exponents and the scaling laws.

2.4 Physico-chemical Applications of SFC

As already mentioned above some useful physico-chemical information can be obtained from SFC experiments. A review of this field has already been presented elsewhere [7] (see also [8–16]). In the following only some characteristic examples will be presented and discussed.

2.4.1 Capacity Ratios

As already mentioned in Sect. 1, the capacity ratio k_i' can be obtained from the retention time t_{Ri} of a sample i according to Eq. (1). Since k_i' is proportional to the partition coefficient K_i, the thermodynamic description of K_i can also be used for k_i' e.g. with respect to its dependences on temperature T, pressure p or density ϱ. A weakness of all these approaches, however, is the incompleteness of information concerning the retention mechanisms in the stationary phase; for the assumption of an adsorption mechanism see e.g. [7].

Figure 14 shows the results of some systematic SFC experiments on naphthalene and fluorene using supercritical CO_2 as a mobile phase [7]. According to Fig. 14a, log k_i' (and consequently k_i' itself) decreases with increasing pressure giving evidence for an increasing solvent power of the mobile phase with rising pressure. Of special interest is the temperature dependence: At low pressures k_i' decreases with increasing temperature

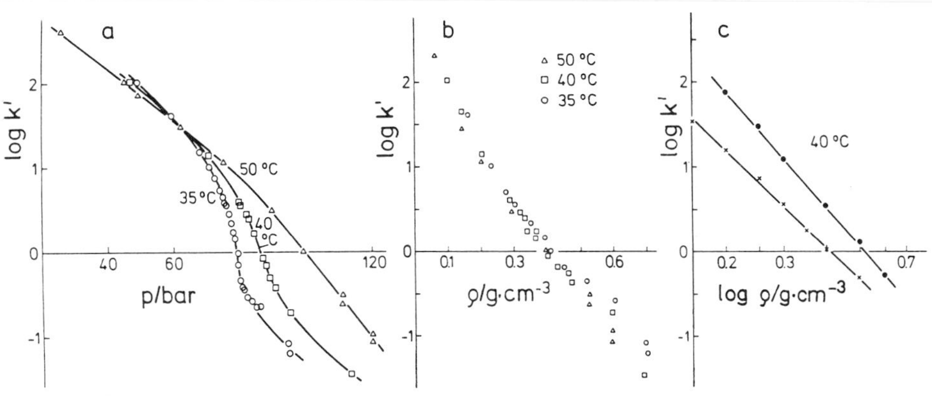

Fig. 14a–c. Capacity ratios k' as a function of pressure p (**a** naphthalene), density ϱ (**b** naphthalene) and $_{10}\log \varrho$ (**c** naphthalene (x), fluorene (o)) (mob = CO_2; stat = column 30.5 cm long, 3.0 mm internal diameter, RP 8, 30–40 μm) (according to van Wasen [19], see also [7])

because here k_i' is proportional to $1/p_i^*$ (where p_i^* = vapor pressure of pure component i), whereas at pressures above about the critical pressure k_i' increases with increasing temperature giving evidence of a lower solubility of component i in the mobile phase under these conditions essentially because of its lower density. Figure 14b and c demonstrate that the dominant parameter in SFC is density: The $\log k_i'(\varrho)$ curve nearly represents a straight line independent of temperature (Fig. 14b), $\log k_i'$ plotted against $\log \varrho$ even gives a better linear fit (Fig. 14c).

Figures 14b and c suggest the following correlations for capacity ratios k_i' as a function of density

$$\log k_i' \approx a \cdot (\varrho/\varrho^\circ) + b \tag{11}$$

$$\log k_i' \approx a' \cdot \log (\varrho/\varrho^\circ) + b' \tag{12}$$

where ϱ° is a reference density. It must, however, be kept in mind that these correlations are good approximations over limited density ranges only.

Plots similar to those in Fig. 14 have been obtained for many different samples, some stationary phases (e.g. reversed phase, adsorbent) and several mobile phases (e.g. CO_2, C_2H_6, SF_6, $CClF_3$ etc.) over wide ranges of temperatures and densities [16–32]. Pressure and density gradients have also been applied and correlated [31, 32].

An example that has been obtained recently is shown in Fig. 15 where $\log k_i'$ values are plotted against density (Fig. 15a) and the reciprocal absolute temperature (Fig. 15b) for some selected liquid crystals as samples [21]. These curves will be important for the analysis as well as the separation or purification of liquid crystals by analytical and preparative SFC. They demonstrate a well-known tendency in SFC: With increasing density

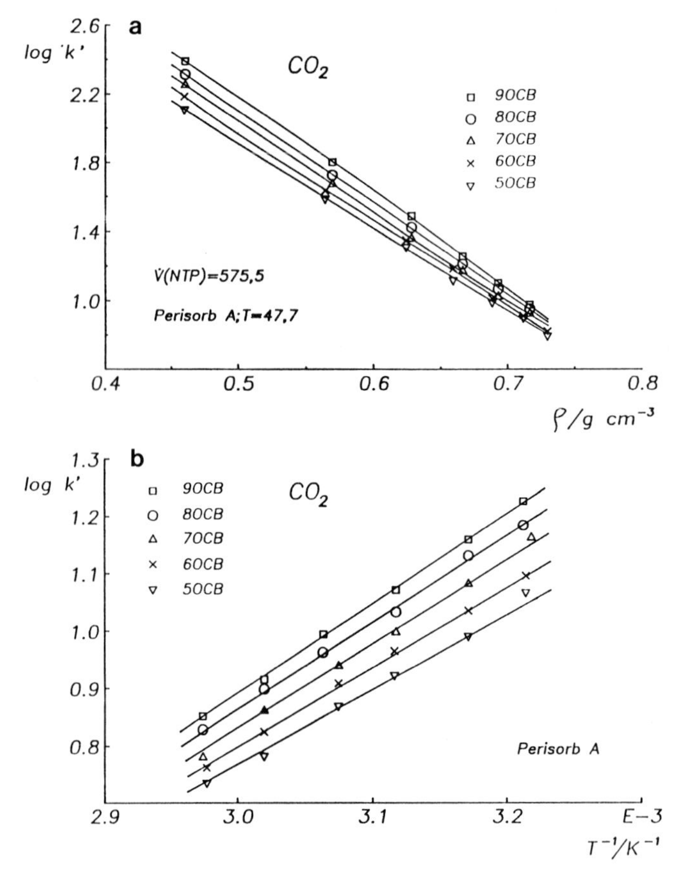

Fig. 15a, b. Capacity ratios k′ of some selected liquid crystals (nOCB = 4-*n*-alkoxy-4′-cyano-biphenyl) as a function of density ϱ (**a** T = 321 K) and temperature T (**b** ϱ = const = 0.70 g·cm^{-3}) (mob = CO_2; stat = column 25.8 cm long, 4.5 mm internal diameter, Perisorb A, 30–40 μm) (according to Jacobs [21])

of the mobile phase, capacity ratios (and therefore retention times) decrease but the selectivities characterized by the quotient k'_i/k'_j also decrease and vice versa. Therefore an optimum between retention time and selectivity has to be found in an application (for the SFC analysis of liquid-crystal mixtures see [51]).

2.4.2 Diffusion Coefficients

As already mentioned in Sect. 1 information with respect to transport properties can be obtained from the peak profile. Binary diffusion coefficients D_{12}^{∞} at infinite dilution can be determined from SFC experiments according to the so-called peak-broadening method using a long void column from the following relation for the height equivalent of a theoretical plate HETP

$$\mathrm{HETP} = \frac{1}{n} \equiv \frac{\sigma^2(z)}{l} \approx \frac{2D_{12}}{\bar{u}} + \frac{r_0^2 \cdot \bar{u}}{24 D_{12}^{\infty}} \tag{13}$$

where n is the efficiency, $\sigma^2(z)$ the peak variance in length units, $\bar{u}$ the mean velocity of the mobile phase, r_0 the inner radius and l the length of the column; the pressure drop along the column is small. For details see e.g. [33].

D_{12}^{∞} values have been determined for many substances in supercritical CO_2, C_2H_6, $CClF_3$, SF_6 etc. They are normally of the order of 10^8 $m^2 \cdot s^{-1}$ and thus much higher than those in liquid solvents. This is of considerable importance for mass transfer in SFE and SFC e.g. concerning HETP values. A characteristic example is shown in Fig. 16 where D_{12}^{∞} values for some fatty acids as a function of ϱ are compared with those of squalene [35]. The data can be qualitatively understood from the fact that with an increasing number of double bonds the molecules become stiffer and more bulky and as a consequence D_{12}^{∞} decreases. For more data and a detailed discussion see e.g. [26, 28, 29, 33–35].

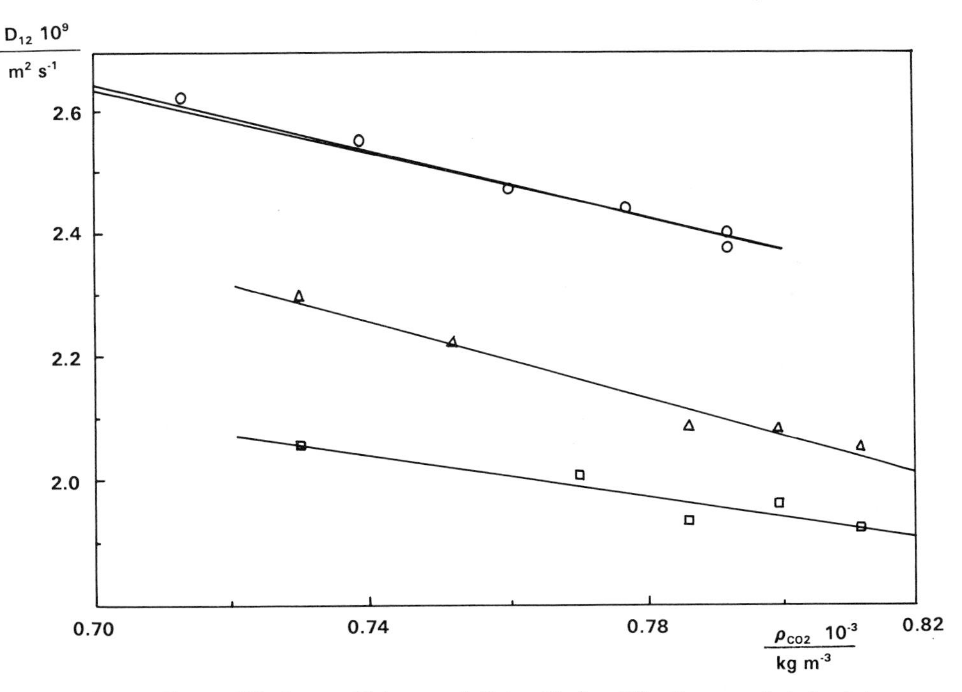

Fig. 16. Binary diffusion coefficients at infinite dilution D_{12}^{∞} of some selected substances (stearic acid (◇), oleic acid (○), linoleic acid (△), squalene (= 2,6,10,15,19,23-hexamethyl-2, 6,10,14,18,22-tetracosahexaene) (□) in supercritical CO_2 as a function of density ϱ at 314 K (mob = CO_2, void column 46 m long, 0.41 mm internal diameter, flow velocity ca 0.5 $cm \cdot s^{-1}$, UV-detector) (according to Dahmen et al. [35])

References

1. Schneider GM (1966) Ber Bunsenges Phys Chem 70:497
2. Schneider GM (1970) Adv Chem Phys 17:1
3. Schneider GM (1973) In: Franks F (ed) Water – A Comprehensive Treatise, Vol 2. Plenum Press, New York, Chap 6, p 381
4. Schneider GM (1975) In: Le Neindre B, Vodar B (eds) Experimental Thermodynamics, Vol II. Butterworth, Chap 16, Part 2
5. Schneider GM (1978) Angew Chem 90:762; (1978) Angew Chem Int Ed Engl 17:716
6. Schneider GM (1978) In: McGlashan ML (ed) Chemical Thermodynamics, Vol 2. A Specialist Periodical Report, London, Chap 4, p 105
7. Van Wasen U, Swaid I, Schneider GM (1980) Angew Chem 92:585; (1980) Angew Chem Int Ed Engl 19:575
8. Schneider GM (1981) Rheinisch-Westfälische Akademie der Wissenschaften, Lectures N301. Westdeutscher Verlag, p 7
9. Schneider GM (1983) Fluid Phase Equil 10:141
10. Schneider GM (1983) Pure & Appl Chem 55:479
11. Schneider GM (1984) Ber Bunsenges Phys Chem 88:841
12. Schneider GM (1985) Thermochimica Acta 88:17
13. Deiters UK, Schneider GM (1986) Fluid Phase Equil 29:145
14. Schneider GM (1987) In: Hirata M, Ishikawa T (Eds) The Theory and Practice in Supercritical Fluid Technology. Chap 1. NTS, Tokyo
15. Schneider GM, Ellert J, Haarhaus U, Hölscher IF, Katzenski-Ohling G, Kopner A, Kulka J, Nickel D, Rübesamen J, Wilsch A (1987) Pure & Appl Chem 59:1115
16. Schneider GM (1988) In: Perrut M (ed) Proceedings of the International Symposium on Supercritical Fluids. October 17–19, Nice (France), p 1
17. Schneider GM, Stahl E, Wilke G (1978) (eds) Extraction with Supercritical Gases. Verlag Chemie, Weinheim
18. Bartmann D (1972) Dissertation. University of Bochum, FRG
19. Van Wasen W (1978) Dissertation. University of Bochum, FRG
20. Wilsch A (1985) Dissertation. University of Bochum, FRG
21. Jacobs K-H (1990) Dissertation. University of Bochum, FRG
22. Bartmann D, Schneider GM (1970) Chem Ing Tech 42:702
23. Bartmann D (1972) Ber Bunsenges Phys Chem 76:336
24. Bartmann D, Schneider GM (1973) J Chromatogr 83:135
25. Van Wasen U, Schneider GM (1975) Chromatographia 8:274
26. Swaid I, Schneider GM (1979) Ber Bunsenges Phys Chem 83:969
27. Van Wasen U, Schneider GM (1980) J Phys Chem 84:229
28. Feist R, Schneider GM (1982) Sep Sci Techn 17:261
29. Wilsch A, Feist R, Schneider GM (1983) Fluid Phase Equil 10:299
30. Wilsch A, Schneider GM (1983) Fresenius Z Anal Chem 316:265
31. Wilsch A, Schneider GM (1986) J Chromatogr 357:239
32. Linnemann KH, Wilsch A, Schneider GM (1986) J Chromatogr 369:39
33. Kopner A, Hamm A, Ellert J, Feist R, Schneider GM (1987) Chem Eng Sci 42:2213
34. Dahmen N, Dülberg A, Schneider GM (1970) Ber Bunsenges Phys Chem 94:384, 710
35. Dahmen N, Kordikowski A, Schneider GM (1990) J Chromatogr 505:169
36. Kulka J (1990) Dissertation. University of Bochum, FRG
37. Spee M (1990) Dissertation. University of Bochum, FRG
38. Hölscher IF, Spee M, Schneider GM (1989) Fluid Phase Equil 49:103
39. Friedrich J, Schneider GM (1989) J Chem Thermodynamics 21:307
40. Deiters UK, Pegg IL (1989) J Chem Phys 90:6632
41. Deiters UK, Swaid I (1984) Ber Bunsenges Phys Chem 88:791
42. Deiters UK (1985) Fluid Phase Equilib 20:275

43. IUPAC: Carbon Dioxide (1976) International Thermodynamic Tables of the Fluid State 3, compiled by Angus S, Armstrong B, de Reuck KM. Pergamon, Oxford
44. Koningsveld R, Kleintjens LA, Diepen GAM (1984) Ber Bunsenges Phys Chem 99:848
45. Katzenski-Ohling G, Schneider GM (1987) Fluid Phase Equil 34:273
46. Michels A, Kleerekoper I (1936) Physica 6:586
47. Michels A, Tenseldam CA, Overdijk SDJ (1951) Physica 17:781
48. Fornari RE, Alessi P, Kikic I (1980) Fluid Phase Equilib 57:1
49. Hicks CP, Young CL (1975) Chem Rev 75:119
50. Takahashi S, Iwasaki H (1966) Bull Chem Soc Japan 39:2105
51. Schoenmakers PJ, Verhoeven FCCJG, van den Bogaert HM (1986) J Chromatogr 371:121

3 Basic Principles of Analytical Supercritical Fluid Extraction

JERRY W. KING and JOHN E. FRANCE

3.1 The Development of Analytical SFE

The analytical chemist in general employs several procedural steps for the characterization of complex samples. These steps can be classified into three major tasks: sample preparation, analysis, and interpretation of the resultant data. Advances in analytical chemistry have reduced the experimental burden and time required for these task areas; however, to date, sample preparation continues to consume much of the analyst's time in the entire analysis protocol. Improved methods for sample preparation would be welcomed in the analytical laboratory as well as techniques that minimize the use of chemical reagents and their attendant disposal problems. This is particularly true in analytical protocols that require the use of toxic or carcinogenic organic solvents in the sample preparation schemes. The relatively new technique of supercritical fluid extraction (SFE) offers the analyst an alternative for preparing samples prior to analysis, that is rapid and environmentally less hazardous. This chapter will describe the basic principles involved in applying the technique for sample preparation in analytical chemistry.

3.1.1 Sample Preparation in Analytical Chemistry

An interesting review by Majors [1] summarizes the use of various techniques utilized in current sample preparation schemes. Filtration and dilution are by far the most common steps that are employed by the analyst in the preparation of his sample. Classical techniques, such as distillation and liquid-liquid extraction are time consuming and cumbersome to apply in specific cases. Often further refinement of the extract is required through such techniques as size exclusion and adsorption chromatography [2]. These low efficiency fractionation methods are gradually being replaced by miniature solid phase extraction columns. Supercritical fluid extraction is a particularly expeditious technique for the removal of non-polar to moderately polar compounds from a variety of sample matrices. However, as will be shown later, SFE can be used for the solubilization and removal of polar compounds by modifying the supercritical fluid with a cosolvent. For trace analysis, many types of compounds can be processed with SFE, since large analyte solubilities in the extraction fluid are not required.

3.1.2 Utilization of Supercritical Fluids in Analytical-Scale Extractions

Historically, the development of analytical SFE has been associated with a form of chromatography. Early SFC studies employed the principle of SFE in many of the injection devices to crudely fractionate complex mixtures [3, 4]. The research of Stahl in Germany combined SFE with thin-layer chromatography [5] for the fractionation of complex mixtures of natural products [6]. The development of practical SFC systems in the early 1980s saw the addition of micro-extraction devices which facilitate "on-line" SFE with various modes of chromatography.

Recent studies on such instruments have demonstrated that SFE has definitive advantages over conventional extraction methods. SFE has been shown to extract quantitatively environmental toxicants [7–12], pesticides [13–18], and many other compounds [19–21]. In general, extractions performed by SFE require less time than those obtained using a conventional method, such as Soxhlet extraction [8, 9, 11, 14, 19, 22]. In addition, SFE yields results having better precision, than when similar extractions are performed by the Soxhlet method [9, 11], even down to parts-per-billion level of the analyte in the sample matrix.

3.1.3 Features of Analytical SFE

Supercritical fluid extraction is a technique that employs a fluid phase having intermediate properties between a gas and liquid, to effect the solubilization of solutes. The advantages that are gained by employing SFE can be traced to the unique physical properties that these fluids possess. Compared to liquid solvents, supercritical fluids have lower viscosities and higher diffusivities, thus allowing more efficient mass transfer of solutes from sample matrices [23]. Another advantage of supercritical fluids is that their solvent power can be adjusted through mechanical compression of the extraction fluid. This feature not only permits selective extraction to be accomplished, but allows the concentration of analytes after extraction, free from any contaminating solvent. Proper choice of the extraction fluid will also allow the analyst to conduct the extraction at low temperatures, a feature which makes SFE particularly amenable to the treatment of thermally-labile substances.

The above definition encompasses the use of these fluids in the field of chemical engineering, as well as their use in analytical chemistry. It is worth noting the similarities and differences that exist in the application of SFE to these two diverse technical areas. The use of SFE in modern process engineering applications was initiated in Germany during the late 1960s to the early 1970s [24] and are well documented in the patent literature [23]. These early engineering studies showed that SFE was a viable alternative to conventional distillation and solvent extraction processes and permitted the processing of substrates whose extraction could be adversely affected by high temperatures and the presence of solvent residuals. It is important in engineering applications of SFE to maximize the yield of the extract using a minimal expenditure of energy [25]. In addition, care must be taken to minimize the extraction pressure since this increases the costs associated with the construction of plants designed to operate at high pressures. Engineers must also be concerned with the conservation of the processing fluid, hence the recycle mode of SFE is commonly employed in many industrial separation schemes.

For the application of analytical SFE, some of the above constraints are removed. For example, the size of the sample or concentration of the target analyte for analytical purposes is usually much smaller than in the engineering case. This has two practical implications in analytical SFE. For one, the quantity of extraction gas required is considerably less than in large scale SFE, therefore an inexpensive and non-toxic fluid can be decompressed into the atmosphere after use. Secondly, for extractions involving trace quantities of analyte, much lower extraction pressures can be utilized, since large, finite solute solubilities in the supercritical fluid are not required. However, for certain analytical applications of SFE, the attainment of maximum solubility may be desired [26]; therefore higher fluid densities are needed in order to shorten the time of the extraction.

3.2 Fluid Properties in SFE

The properties of a supercritical fluid are of paramount importance when considering the selection of a fluid as an extracting agent. The analyst should strive to select a fluid that exhibits the best compromise in solubilizing the solutes of interest as well as the mass transfer characteristics required to rapidly effect the extraction of the analytes. Optimization of these two factors will assure a high flux rate of the analyte into the extracting medium, thereby saving consumption of fluid, while assuring rapid sample processing. Giddings [27] has suggested that the solvent properties of a supercritical fluid could be partitioned into a "state effect", described by the variation in the physical properties of fluid as a function of compression, and a "chemical effect" which is related to the static physical constants of the gas (fluid). We shall examine the impact of these parameters on the performance of specific fluids in the following sections.

3.2.1 Selection of the Supercritical Fluid

Two parameters which are of prime importance when considering the selection of a supercritical fluid are the critical pressure and temperature. The

Table 1. Critical constants for some common SFE solvents

Compound	Critical temperature (K)	Critical pressure (MPa)	Critical density (g/cc)
Ethylene	283.0	5.12	0.23
Carbon dioxide	304.1	7.39	0.47
Nitrous oxide	309.6	7.26	0.46
Propane	369.8	4.26	0.22
Sulfur hexafluoride	318.8	3.76	0.75
Methanol	513.4	7.99	0.27
Water	637.0	22.1	0.32
Ammonia	405.4	11.3	0.24
n-pentane	469.8	3.37	0.23

critical pressure to a first approximation determines the magnitude of fluid's solvent power in the condensed state and therefore can be used as a crude guide to match the fluid with the anticipated polarity of the compounds to be extracted. For example, ethylene has a lower critical pressure than carbon dioxide, as shown in Table 1. Based on this criterion, ethylene would not dissolve a moderately polar solute to the same extent as carbon dioxide. Likewise, fluids which exhibit higher critical pressures than carbon dioxide, are known to solubilize polar moieties at higher concentrations in the fluid phase than SC-CO_2.

The critical temperature of the fluid exerts its influence in both a theoretical and practical manner. From a practical perspective, one should consider the effect of extraction temperature on the thermal stability of the target analyte. Fluids which are characterized by high critical temperatures will require elevated extraction temperatures in order to affect extraction in the supercritical state. Conversely, fluids having sub-ambient critical temperatures may require cooling of the extractor circuit in order to promote densification of the extraction medium. Theoretically, the maximum extraction gas density is obtained by selecting an extraction temperature which is close to the critical temperature of the chosen fluid. This effect can be demonstrated by comparing the relative densities of carbon dioxide with those of nitrous oxide over the pressure range encompassing their respective critical temperatures (see Table 1) at an extraction temperature of 40 °C (313 K). At the selected extraction temperature, nitrous oxide exhibits a larger increase in density with pressure than CO_2, since its critical temperature is closer to the chosen extraction temperature [28].

The use of specific fluids can also enhance the solubility of a particular class of analytes during SFE. For example, the solubility of certain opium alkaloids in fluoroform is much larger than their recorded solubilities in SC-CO_2 over the same range of extraction pressure and temperature [29]. This trend is due to the tendency of fluoroform to exhibit a specific propensity for hydrogen bonding with alkaloid moieties in the supercritical fluid phase. Evidence for specific adduction in supercritical fluid state has also been presented by King et al. [30], based on the computation of complexation constants from virial coefficient data for SC-CO_2-alcohol mixtures at moderate compression.

3.2.2 Unique Properties of Supercritical Fluid Carbon Dioxide

Inspection of Table 1 reveals that carbon dioxide is unique among the candidate fluids for effecting supercritical fluid extraction. Its critical temperature of 31 °C is close to room temperature, thereby permitting extractions to be carried out at low temperatures on thermally labile compounds. Modest compression of CO_2 produces a substantial change in its fluid density due to the high non-ideality exhibited by this fluid. Carbon dioxide is also non-flammable and odorless; properties which facilitate its use in a laboratory environment. Fluid carbon dioxide is also relatively inexpensive and available in satisfactory quantities. With proper ventilation, it represents little harm to the analyst.

The solubility of solutes in carbon dioxide in both the fluid and liquid phase have received considerable study in the past two decades [31]. The classical study of Francis in 1954 [32] is also worth consulting, since it qualitatively describes the solubility trends for 261 compounds in near critical CO_2. Even though supercritical carbon dioxide preferentially extracts non-polar compounds, it can exhibit an induced dipole moment [33], which enhances the extraction of moderately polar solutes into the fluid phase. Generalized solubility rules, formulated by Stahl [5], support the above observations, and indicate that the introduction of polar functional groups into the molecular structure of a compound results in a substantial reduction in solute solubility in the SC-CO_2 phase. For trace analysis, such a reduction in solubility may not be deleterious, provided that the solubility level of a compound is adequate for analytical SFE and detection.

In general, SC-CO_2 is an excellent solvent for the extraction of lipophilic solutes from a variety of sample matrices. Under the proper extraction conditions, an appreciable amount of lipid material can be solubilized in SC-CO_2, which in certain specific applications of SFE can be an advantage or disadvantage. In micro-analytical coupled SFE, the high solubility of lipids can result in the extraction of excessive amounts of solute, which may overload chromatographic columns and impact on peak resolution. In other specific cases, the high solubility for these non-polar solutes may permit the selective and easy isolation of a lipid phase associated with a sample matrix.

The basis of the latter example is illustrated in Fig. 1 where the solubility of soybean oil triglycerides is plotted as a function of extraction pressure. In this case, triglyceride solubility is a function of both extraction pressure and temperature. In the case of the 80 °C isotherm, there appears to be a minimum "threshold pressure" which must be attained to solubilize a

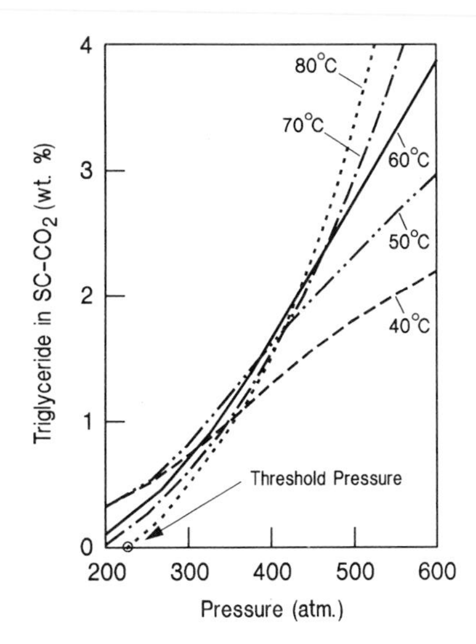

Fig. 1. Solubility of soybean oil triglycerides in $SC\text{-}CO_2$ as a function of pressure and temperature

"measurable" amount of lipid. Another feature in the solubility trends for triglycerides is the inversion in solubility at a particular temperature as extraction pressure is increased. This pressure interval over which the inversion in solubility occurs is called the "crossover" region, and its occurrence permits the partial fractionation of solutes [34, 35]. Beyond the crossover region, the lipid solubility increases with pressure, reaching a solubility maximum at very high pressures. The appreciable increase in lipid solubility can reach values in excess of 25% by weight, thereby permitting the rapid isolation of lipids from large samples.

3.2.3 The Use of Cosolvents in SFE

In certain cases, it becomes desirable to add a cosolvent to a supercritical fluid to enhance the solubility of an analyte in the extracting medium. Such cosolvents, also called entrainers or moderators, are usually organic solvents, that are added to the source of compressed fluid before the pump or compressor, or alternatively, to the extraction gas after it is compressed, using a high pressure liquid pump. The addition of a cosolvent to a supercritical fluid not only enhances the solubility of a analyte, but in specific cases will improve the separation factor between solutes as they are selectively partitioned into the supercritical fluid phase. Maximum separation factors are usually achieved at low solute levels in the supercritical fluid phase, a condition that is more amenable to analytical SFE than to engineering-scale SFE. The function of a cosolvent in a large scale SFE is primarily to increase the solubility of non-volatile components in the compressed gas phase and to facilitate solute separation after extraction, without resorting to a reduction in the processing pressure [36].

To date, there is no theory that can satisfactorily predict the effect of cosolvent addition on the solubility of solutes in a supercritical fluid phase. The effect of a particular cosolvent must largely be determined by experiment observation. Recently, however, a thermodynamic rationale for cosolvent selection has been proposed by Johnston and coworkers [37]. To a first approximation, the volatility enhancement of solutes into supercritical fluids is controlled by the densification of the extraction fluid; cosolvents modify the polarity of the extracting phase, and can thus improve the separation factor between solutes that differ in the number and type of functional groups in their molecular structure.

Wong and Johnston [38] have shown for similar types of solutes such as sterols, that the relative solubilities in SC-CO_2 are primarily determined by the respective vapor pressures of the sterols. The addition of a polar organic cosolvent to the SC-CO_2 in this study was found to enhance the solubility of the sterols by up to two orders of magnitude and to provide a significant enhancement of one sterol moiety over another in the supercritical fluid medium. Greater enhancements have been reported when using cosolvents which hydrogen-bond with specific solutes as demonstrated by the 620% increase in solubility reported for 2-aminobenzoic acid upon the addition of 3.5 mole% of methanol to SC-CO_2 [39].

The use of cosolvents in SFE requires that the analyst choose his extraction conditions judiciously. For example, the addition of a cosolvent to the fluid phase will change the critical point of the mixture from the one recorded for the pure supercritical fluid. Hence, it is important to recognize the magnitude of this change so as to adjust the experimental parameters commensurate with conducting an extraction in the one phase region. Likewise, the solubility of the cosolvent in the supercritical fluid is determined by the extraction temperature and pressure, therefore the quantity of the cosolvent that can be added to the fluid phase must be regulated.

Cosolvent addition to supercritical fluids may also provide some additional benefits that can improve the extraction. For example, it has been demonstrated that the addition of water to the fluid medium can change the morphology of the substrate that is being extracted and result in an improved extraction flux of a particular component, e.g. caffeine from coffee [24]. Cosolvents may also aid in the desorption of analytes from highly adsorptive sample matrices by displacing the analyte from the surface as opposed to increasing its solubility in the supercritical fluid phase. The molecular mechanism appears to be one of competitive adsorption between the cosolvent, supercritical fluid, and adsorbed analyte on the surface of the sample [12].

In recent years, very specific and novel agents have been added to various supercritical fluids to provide both enhanced and selective extraction of particular solutes. The use of an ion pairing agent such as tetrabutylammonium hydroxide has been cited for the removal of polar drugs from aqueous solution [40]. Non-polar supercritical fluids containing reverse micelles have been used to solubilize high molecular weight proteins and inorganic salts in the compressed gas phase. For example, the amino acid, tryptophan, can be solubilized at levels 100 times more than that obtained with pure supercritical ethane by the additional of the anionic surfactant, sodium di-2-ethylhexyl sulfosuccinate, to the fluid phase [41]. Such specificity creates many interesting possibilities for the analyst over and above the results that have already been reported using conventional SFE.

3.3 Optimizing Experimental Conditions for Analytical SFE

In the practice of analytical chemistry, many analysts utilize an empirical approach to arrive at the best experimental or analysis conditions. Such an approach is made, in part, because of time limitations that are placed on the analyst and partly due to the lack of theoretical guidelines which can be utilized when analyzing complex samples. Many theoretical approaches have been developed for predicting the solubility of solutes in supercritical fluid media [42] but they are of limited value to the practicing analyst because of computational time required and the lack of physical property data on the target analytes, solvents, etc. Complicating the problem is the influence of the sample matrix which may have a synergistic or retardative effect on the recovery of solutes.

3.3.1 Objectives of the Extraction

To chose the optimal conditions for performing SFE, the analyst must define his analytical objective, since for many applications, SFE is not a highly discriminative technique. In fact, SFE rarely allows the isolation of one specific analyte to the exclusion of other co-extracted moieties. A crude form of fractionation can be accomplished by SFE [43] by simply varying either the extraction pressure or temperature. Such a method yields positive results only when there are significant differences in the molecular weight or polarity between the components that are being extracted.

An example of this approach is illustrated in Fig. 2 where SC-CO_2 has been used to fractionate the components in grapefruit oil during an on-line SFE/SFC extraction. Utilizing a fluid density of 0.18 g/cm^3 for the extraction, allows the isolation of specific components from the oil matrix. This low-pressure or density "skimming" of selected compounds can be used to advantage to analyze specific components in a complex matrix [14]. The use of higher extraction densities removes additional compounds; however, inspection of Fig. 2 reveals there are common components that are extracted at each fluid density. The lack of selective extraction can tax the chromato-

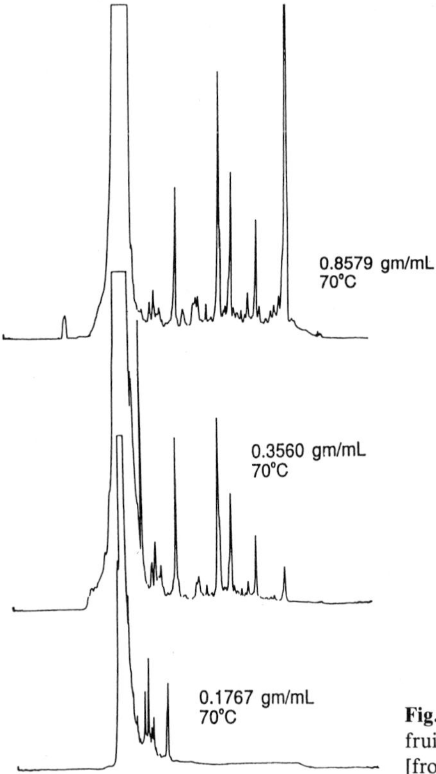

Fig. 2. SFE-SFC profiles of cold pressed grapefruit oil obtained at different extraction densities [from Ref. 82]

graphic resolving capabilities of an on-line extraction system, particularly when the target analytes are minor components in the SFE.

Total or complete extraction of the sample matrix by SFE requires that the extraction time be optimized [44]. Fractionation of a mixture by SFE shows not only a dependence on the experimental pressure and temperature, but time as well. The observed molecular discrimination that occurs as a function of time is often molecular weight dependent. Hence, for an integral off-line extraction, the initial samples taken on a sequential basis will be enriched in lower molecular weight components. The time required to complete a total extraction of the sample will be dependent on the mass of the sample taken for extraction and the level of extractable material (by SFE) in the sample. Total SFE of lipid phases from samples by SC-CO_2 has been reported and can be accomplished quite easily by off-line SFE [26].

The above citations are but two examples of the versatility of SFE. To apply SFE properly, the analyst must become familiar with several critical parameters that have impact on the extraction. These are discussed at length in the next section.

3.3.2 Critical Parameters Pertinent to SFE

There are several parameters which are of key importance in conducting an extraction by SFE. These are 1) the pressure (or density) at which the desired solute becomes miscible with the supercritical fluid phase, 2) the condition for attaining maximum solute solubility in the extraction fluid, and if available, 3) the physical and chemical properties of the solutes that are to be extracted. In general, knowledge of the miscibility pressure and solubility maxima for a given solute/fluid system will define the potential range of pressures for fractionating the extracted components [13].

The miscibility pressure for a given solute-solvent pair remains a somewhat ill-defined concept, since it depends on the technique that is used to measure the solute's solubility in the supercritical fluid. Giddings [28] defined the concept of the "threshold pressure" based on the ability of a flame ionization detector to detect a solute in a flowing stream of supercritical fluid. King [45] has noted that it is possible to employ a range of techniques for determining a solute's threshold pressure in a supercritical fluid, that vary in sensitivity over a range of 10^9! For the analyst, it is most important to have a knowledge of the relative miscibility pressures of different compounds, since their relative magnitudes will determine the feasibility of separating components based on this principle. Unfortunately, many compounds exhibit small differences in their respective miscibility pressures, so that selective SFE of individual compounds proves difficult in practice, due to the high precision required in regulating the extraction pressure and the commensurate small solute solubilities that exist under these conditions.

Knowledge of the pressure required for achieving maximum solubility of the solute in the supercritical fluid phase has several important implications in analytical SFE. Extractions conducted under these conditions can shorten the extraction time and also permit the processing of larger samples for analysis. Unfortunately, there is usually a loss in extraction selectivity under these conditions and the compression requirements can be quite high. For example, the removal of lipid phases from natural products is best affected at pressures in excess of 69 MPa, where infinite miscibility of triglycerides in SC-SO_2 is attained [46]. Extractions conducted at these conditions can remove grams of lipid from a sample within fifteen minutes [26].

The physical properties of the solute can also play a role in SFE. Particularly germane are the compound's melting point and vapor pressure. Threshold pressures have been shown to be dependent on the melting point of a compound and it has been observed that compounds with melting points in excess of 350 °C are not readily solubilized in dense CO_2 [47]. SFE is also more easily affected when the extraction is conducted at a temperature above the compound's melting point, since the solid's cohesional energy is reduced. The effect of a solute's vapor pressure on the enhancement factor in SFE has been described theoretically [48] and it has been generally observed, that beyond a certain pressure, a compound's solubility in a supercritical fluid increases with increasing temperature [49]. This latter trend is due to the substantial reduction in the solute's cohesive energy density with increasing extraction temperature which outweighs the commensurate loss of solvent power in the extracting fluid.

3.3.3 A Theoretical Approach for Optimizing SFE

As noted previously, analysts are prone to ignore theoretical approaches for optimizing SFE, due to the lack of time or complexity of the theory. In the course of our research, we have developed a relative simple method, based on combining the solubility parameter theory with the Flory-Huggins interaction parameter concept, which explains many of the salient features of SFE. The data required by the above theory consists of fluid and solute critical or reduced property data and solubility parameters. Such data, if unavailable, can be estimated from corresponding states theory [50], group contribution methods [51], or nomographs [52].

The key equations utilized in this approach are

$$\chi = \chi_H + \chi_S = \frac{\bar{V}_1}{RT}(\delta_1 - \delta_2)^2 + \chi_S \tag{1}$$

where χ is the total interaction parameter χ_H, χ_S are the enthalpic and entropic interaction parameters, respectively; δ_1, δ_2 are the solubility parameters of the supercritical fluid and solute, respectively, and $\bar{V}$, is the molar volume of the fluid. The parameters δ_1, δ_2 and $\bar{V}_1$ are dependent on pressure and temperature and this factor must be taken into account when computing their values. The solubility parameter of the supercritical fluid, δ_1, is calculated by the method of Giddings [27] as

$$\delta_1 = 1.25\,P_c^{1/2}(\varrho_r/\varrho_{r_1\,liquid}) \tag{2}$$

where P_c is the critical pressure of the fluid, ϱ_r is the reduced density of the supercritical fluid, and $\varrho_{r_1\,liquid}$ is the reduced density of the near-liquid fluid at conditions approaching infinite compression.

Equation 1 can be rearranged to reveal its functional dependence on pressure. In this form, assuming a constant value for χ_S, plots of χ_H versus pressure are hyperbolic, with the a minimum occurring at a pressure when δ_1 equals δ_2. At this condition, maximum solubility of the solute occurs in the supercritical fluid. A common misconception in applying the above theory to SFE is that the extraction conditions must be chosen to achieve maximum solute solubility in the extracting fluid. However, for certain applications of SFE, such as the extraction of trace quantities of analyte, much lower pressures and δ_1 will suffice. In addition, it is not necessary for the respective solubility parameters of the fluid and solute, δ_1 and δ_2, to match, to achieve an effective SFE. Figure 3 illustrates this principle with a solubility parameter scale encompassing the values for a number of polar

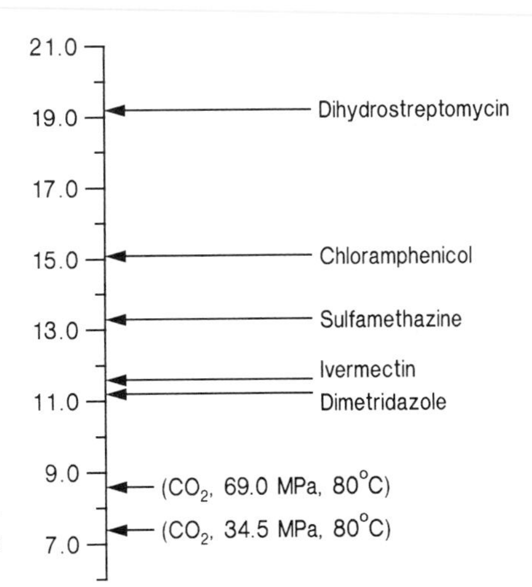

Fig. 3. Solubility parameter scale for SC-CO_2 at specified pressures and assorted drug analytes

compounds (drugs) and SC-CO_2 at two levels of compression. The obvious mismatch in solubility parameters of the drugs and extracting fluid does not mean that trace quantities of drugs, such as sulfamethazine or chloramphenicol, cannot be extracted into SC-CO_2. However, significant differences in δ_1 and δ_2 indicate a low solubility of the solute in the extracting fluid [13], and this may limit the quantitative recovery of such moieties in SC-CO_2.

Assessment of the miscibility pressure of the solute in the supercritical fluid phase can be approximated by employing the Flory [53] critical interaction parameter concept. Here, the critical interaction parameter, χ_c, is given by Eq. 3 as

$$\chi_c = (I + x^{1/2})^2/2x \tag{3}$$

where $x = \bar{V}_2/\bar{V}_1$ and $\bar{V}_2$ is the molar volume of the solute. Since χ_c is a function of $\bar{V}_1$ and $\bar{V}_2$, χ_c will have a weak dependence on pressure. Plots of χ and χ_c as a function of fluid pressure will show a common intercept, whose value on the pressure axis corresponds to the solute's miscibility pressure in the supercritical fluid. The above approach has been shown to accurately predict the miscibility of the pesticide, DDT, with SC-CO_2 [13] and the pressure range which is applicable for the fractionation of oligomers in SFC.

3.4 The Relevance of SFC-Derived Data to Analytical SFE

As a technique, supercritical fluid chromatography can be used independently of SFE, as attested in the other chapters of this book. However, in many cases, SFC can assist in the development of an analytical SFE method. The relevance of SFC-derived data goes beyond its use as a characterization tool for SFE-derived extracts, either on an analytical scale [40, 54–63] or for monitoring pilot plant SFE processes [64]. Two generic areas of application are worth noting: (1) the utilization of SFC measurements to derive data relevant to analytical SFE, and (2) the implications of such measurements to the analytical chemist in SFE method development.

3.4.1 Relevant Measurements by SFC

The application of SFC for the measurement of physicochemical data has been documented in several review articles over the past decade [65, 66]. In this section, we shall primarily be concerned with noting the relevance of such data to analytical SFE. Table 2 tabulates some of the parameters which can be derived from various SFC experiments.

Data such as diffusion and virial coefficients are derived from solute peak broadening experiments or the pressure dependence of retention constants, respectively [67, 68]. Although these parameters are of value in optimizing a SFE, they provide no immediate information to the analyst that is likely to help in solving the analytical problem at hand. The determination of sorption isotherms or the measurement of solute solubilities in supercritical fluids by chromatographic methods require the construction of specialized apparatus [69] and employ different forms of chromatography [45]. Such experiments require excessive time and knowledge that may not be available to the analytical chemist.

An example of a useful and rapid SFC experiment for the analytical chemist is depicted in Fig. 4, where the miscibility pressure as a function of temperature has been determined for the pesticide, malathion, in SC-CO_2. The data were taken using a supercritical fluid chromatograph equipped with a nitrogen-phosphorus detector, which permitted the detection of trace levels of the organophosphorus pesticide in the column eluent. The instrument was operated at various combinations of pressure and temperature that were sufficient to just elute and detect a trace quantity of the injected pesticide. Therefore, at a temperature of 40 °C, a CO_2 pressure of approximately 75 atmospheres was required to solubilize a small quantity of

Table 2. Parameters derived from SFC experiments

Diffusion coefficients
Sorption isotherms
Phase distribution constants
Solubility measurements
Critical loci
Solute partial molar volumes
Virial coefficients

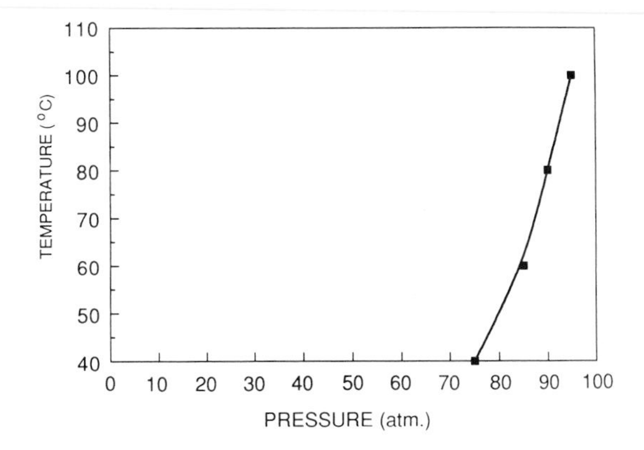

Fig. 4. Miscibility pressure for malathion in SC-CO_2 as a function of temperature and pressure as determined by SFC-nitrogen/phosphorus detector

malathion into the supercritical fluid phase. Note that an increase in temperature required a higher fluid pressure in order to initially solubilize the pesticide into the dense carbon dioxide phase. The data derived from the above experiment provides the analyst with information concerning the minimum extraction pressure required to extract trace levels of the pesticide.

Similarly, King and coworkers [70] have utilized elution SFC to ascertain the ease of extraction of soybean oil volatiles from various sorbents. In these experiments, simple columns were constructed containing the sorbent material and inserted into the SFC. Elution pulse chromatographic experiments were performed at various extraction conditions to determine the pressure and temperature required to desorb the solutes from the adsorbent column. The derived data proved of value not only in choosing the proper sorbent and conditions for stripping the volatiles from a large scale SFE process stream, but in the design of a side stream sampling device to collect the solutes under elevated pressures and temperatures.

3.4.2 Implications for SFE

Other uses for SFC in the development of SFE methods are more indirect then those cited above, but just as beneficial to the analytical chemist. For example, the SFC literature is an excellent source for identifying compounds that are most likely to be amenable to SFE [71–73]. To illustrate this point, the work of Fields and Grolimund [74] is worth citing, since it demonstrates which amine solutes can be solvated in SC-CO_2 using capillary SFC. Similar experiments can be performed quite easily by analyst, since they provide instant verification on the extractability of a particular compound. Obviously, if the compound in question does not chromatograph successfully, then the analyst should consider modifying the conditions for SFC, or perhaps explore the possibility of choosing an alternative extraction fluid or cosolvent.

SFC retention data has also been shown to have a predictive value in determining the correct cosolvent concentration that is required for the extraction of thermally labile herbicides from a variety of sample matrices [75]. In this case, elution order was found to correlate with relative extraction efficiency of the herbicides from various soils. Examination of other chromatographically-derived parameters, such as solute capacity factor [76], should aid the analyst in the selection of extraction conditions or for the choice of cosolvents [77]. It should be recognized however that solute chromatographic mobility through columns may not always mimic the environment under which the SFE is performed, and the analyst should exert caution in correlating the results from the two techniques.

3.5 The Practice of Analytical SFE

The practice of SFE requires that the analyst construct or purchase suitable equipment for conducting the extraction and establish the proper experimental conditions required for the development of the analytical method. A key question facing the analyst will be whether to conduct the extraction in the "off-line" mode, or to combine the extraction step with another analytical technique, thereby facilitating a "on-line" mode of extraction. Utilization of the off-line mode of SFE offers many advantages to the analyst who is inexperienced in SFE. On-line SFE requires that the analyst understand and control more then one technique simultaneously, and therefore may not be the best starting point for the novice to SFE. In the following sections we shall discuss the equipment and experimental variables that are required to conduct both modes of SFE.

3.5.1 Equipment Requirements

Both off-line and on-line modes of SFE share a common base in the equipment that is used to facilitate the extraction. Generically speaking, an analytical supercritical fluid extractor consists of a source of fluid, a fluid delivery module, an extraction cell, a backpressure regulating device, and a collector for trapping the extract after SFE. The fluid is usually supplied in high pressure gas cylinders, which can be equipped with inductor tubes, if the fluid is to be pumped to into the SFE apparatus. In some cases, the fluid tank pressure has been found to be sufficient for performing SFE of trace levels of toxicants from various sample matrices [14].

There are a plethora of fluid delivery devices ranging from high pressure diaphragm compressors, to gas booster pumps, or reciprocating piston pumps, and syringe pumps. Each device has its own merits, but there are some general principles worth considering in selecting the fluid delivery module. Many of the pumps used for delivery the fluid in analytical SFE are modified high performance liquid chromatography (HPLC) pumps and require an external cooling source to assure liquefaction of the fluid. Such cooling is critical to the performance of plunger-based pumps to avoid cavitation due to the introduction of a two phase fluid mixture at the pump head. Syringe delivery pumps that are used for both SFE and SFC, also require a source of coolant for effective operation. It is well known in engineering-scale SFE, that liquid pumps are thermodynamically more efficient than compressors for delivery of the fluid to the extractor. Delivery of a liquified gas also permits easy blending of cosolvents with the principal fluid. The merits of using compressor technology mainly lie in the increased fluid flow capacity that can be obtained with relatively inexpensive equipment and the elimination of the external cooling requirement cited above for pumping modules. Compressor based fluid delivery devices may also require the use of ballasts to dampen to pneumatic pulse from the compression stroke and its effect on the flow of the fluid.

Extraction calls have been fabricated out of a variety of materials, but most cells consist of a tubular metal cavity with associated compression fittings. Many of the reported on-line SFE studies have incorporated modified HPLC column guard cartridges as sample holders. Such cells have finite lifetimes, principally due to the development of leaks around the sealing components of the cell. Alternative sealing methods exist that are based on high pressure coned-and-threaded connections [26] which yield very long service lifetimes. The size of the cell should be scaled to avoid excessive void volume, which in turn reduces the amount of fluid required for the extraction. Incorporating diffusers at both the entrance and exit of the extraction cell will assure that the fluid stream makes contact with the entire sample during the extraction.

Devices used for regulating the extraction pressure on the extractor cell have ranged from narrow fused silica capillary tubing, to bona fide backpressure regulators, or micrometering valves having adjustable flow orifices. The selection of the device is governed by fluid flow rate desired through the extraction cell and upstream pressure desired in the extractor. The use of a silica capillary as a rate limiting orifice requires that the analyst adjust the flow rate through the extraction cell by varying the length or internal diameter of the capillary tubing. Such a procedure is cumbersome compared to using a micrometering valve or a accurate backpressure regulator in conjunction with a precise fluid delivery pump [26]. Another viable option is to use a micrometering valve for flow control into the extraction cell and a backpressure regulator at the exit of the cell.

The need to isolate the extract after SFE has fostered a number of ingenious collection schemes. A cardinal principle in selecting a collection device is to use a vessel with sufficient volume to insure adequate collection efficiency of the extracted solutes. Utilization of too small a collection volume can result in entrainment of the extracted solutes in the expanding fluid stream. Many investigators have incorporated packing in the collector to induce precipitation or applied cooling to assure that an adequate phase separation takes place. Collector sizes have ranged from flasks containing several hundred milliliters [60] to small, 1–2 milliliter, vials [7]. The Joule-Thomson expansion which accompanies the expansion of the fluid to ambient conditions can be used to advantage when collecting the extracted sample in volatile solvents [78]. The cooling power of the expanding extraction fluid prevents evaporation of the collection solvent which absorbs the target analyte. Collection in enclosed vessels equipped with sampling tubes can be used to advantage when sequential sampling during the extraction is desired [26]. Sorbent trapping, or "accumulators" have also been utilized for collecting extracted analytes and this method may be preferred when the fluid stream contains volatile analytes.

A discussion of the large variety of extractors is beyond the scope of this chapter. Designs range from very simplistic, single sample off-line extractors, to complex "on-line" versions which incorporate syringe pumps for both SFE and SFC and utilize cryocooling or retention gaps with sophisticated valving to transfer the solute between the extraction and chromatography module. A very simple off-line extractor is portrayed in Fig. 5 and will serve as an example to convey how a supercritical fluid extractor is assembled. In this example, liquified carbon dioxide is supplied to a syringe pump through a high pressure cylinder equipped with a inductor tube. The liquified gas is converted to the supercritical fluid state prior to filling

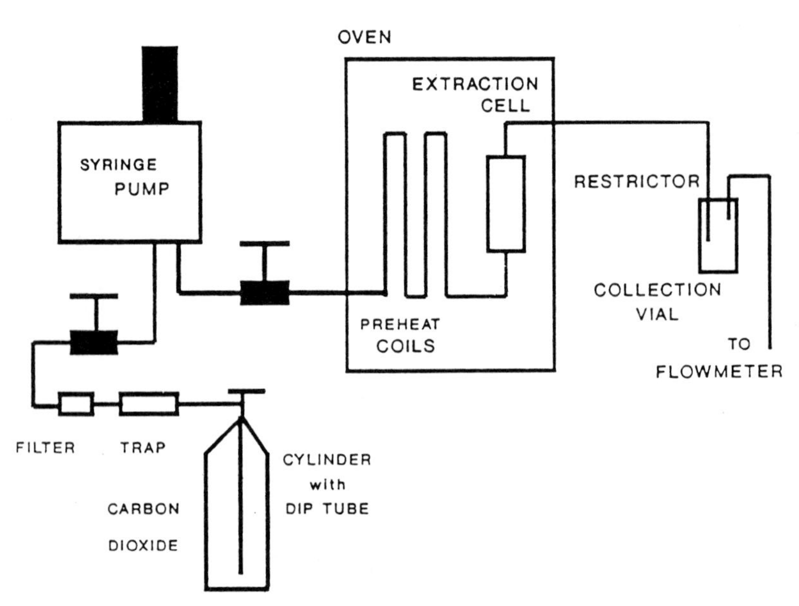

Fig. 5. Schematic of an off-line supercritical fluid extractor

the extraction vessel through the use of a heat exchange coil. The back-pressure regulating device in this case is a narrow bore tube which restricts the flow and increases pressure upstream. Collection of the extracted material is facilitated in a vial. Such a system may also include a flowmeter or gas totalizer to assess fluid flow or passage.

3.5.2 Experimental Considerations in Analytical SFE

The analyst must consider a number of experimental details to achieve a successful SFE. For example, the effect of temperature on the physical state of the supercritical fluid and the resulting phase equilibria is particularly important in executing SFE. It is critical that the SFE system be adequately heat traced so that a single phase is maintained prior to decompression and collection of the extracted solutes. Failure to maintain isothermal conditions in the extraction cell can lead to variable results; however, such conditions are not required throughout the entire system, provided that analyte precipitation is avoided prior to its capture from the fluid stream. It is important to preheat the supercritical fluid prior to its introduction to the extraction chamber. This step can be accomplished through the use of a coiled fluid introduction tube or a more formal heat exchanger arrangement. Likewise, heating of the decompression valves or regulators downstream from the extraction cell is standard practice, in order to assure that precipitation of the extracted analytes does not occur in the valve orifices. Such a situation could cause cessation of fluid flow, due to the cooling effect accompanying the expansion of the supercritical fluid to ambient conditions. Surprisingly, a number of commercial SFE devices have been constructed which violate some of the above criteria.

The materials which comprise extraction equipment are usually checked with respect to their pressure ratings, but frequent physical inspection of the cell and its components should be standard practice. Cell and associated tubing or valves should be inert with respect to the extraction conditions. Sample matrices which contain water should be extracted using rust resistant alloys, since even carbon dioxide can be regarded as a weak acid anhydride. A more critical component are the polymeric materials used as o-rings and seals in valve stems and associated high pressure fittings. Such elastomers may undergo expansion under compression or be solvated by contact with the supercritical fluid media. This phenomena may lead to leaks in the extraction system or extraction of unwanted contaminants into the isolated extract.

Another critical parameter in performing SFE is the design and orientation of the extraction cell. The size of the sample cell should be sufficient to address the analysis problem. As noted previously, many in-line extractors coupled to supercritical fluid or gas chromatographs utilize cells ranging in volume from 1 – 10 milliliters. Cells of this capacity are particularly appropriate for cases where the available sample size is small, however they are of insufficient capacity for other types of analyses. Sizing of the extractor vessel should never be based on the available fluid delivery system, but should be scaled to a level that will allow extractions to be performed on a sample size that accurately reflects the matrix being analyzed. Large extraction cells along with ancillary equipment have been described for use in performing off-line analytical assays [26].

Cell geometries have tended to be cylindrical in deference to the variety of high pressure tubing utilized in constructing extraction cells. Other designs have been reported which permit recirculation of the extracting fluid through the sample matrix [79] or transmission of the supercritical gas through a aqueous matrix [40]. Using identical cells, Andersen et al. [80] have noted differences in extraction efficiency depending on whether the extraction cell was held in a vertical versus horizontal position. In certain cases, a physical inversion of the sample and the extracting fluid may occur within the cell due to change of density of the extracting medium with respect to the sample. Regardless of the extraction cell's spatial arrangement, it is important to have the vessel well packed with the sample of interest and to provide sufficient fluid to assure that the extraction is complete.

The kinetics of a SFE parallels trends found in liquid-liquid extraction. In general, quasi-equilibrium extraction gives way to diffusional-limited extraction kinetics in the latter stage of the SFE. Factors which impact on the rate of completion of the extraction, such as the sample matrix, will be discussed in the next section. From a practical perspective, it is important to ascertain how much fluid is required to complete the SFE. This parameter may be expressed as a volume of fluid at either supercritical fluid or ambient conditions, or more preferably the mass of extraction fluid. Some researchers [81] advocate the use of cell volumes as a measure of the fluid required to complete the extraction. This is a questionable practice since the contents of the cell can be compressed during the course of the extraction, thereby changing the void volume of the cell.

In addition, changes in the flow rate of the fluid can occur during the extraction. These variations in fluid flow can be measured with the aid of a mass flowmeter or flow totalizer unit. The completeness of a off-line extraction can be monitored by taking intermittent samples for analysis. For extraction systems in-line with a form of chromatography, simply running a second extraction will usually suffice to indicate whether the analyte of interest has been completely removed.

The isolation of the extract, either from a on- or off-line SFE system requires that certain precautions be taken. For instance, the temperature chosen for enacting cryofocusing can introduce a bias into the sample that is trapped in the tee or retention gap [82]. This will change the distribution of the components in the resultant extract and hence can be used as a aid in fractionating the total extract. Similarly, sorbent-filled columns can be used to fractionate or isolate specific analytes from the fluid stream after extraction, either at ambient or pressurized conditions. In the former case,

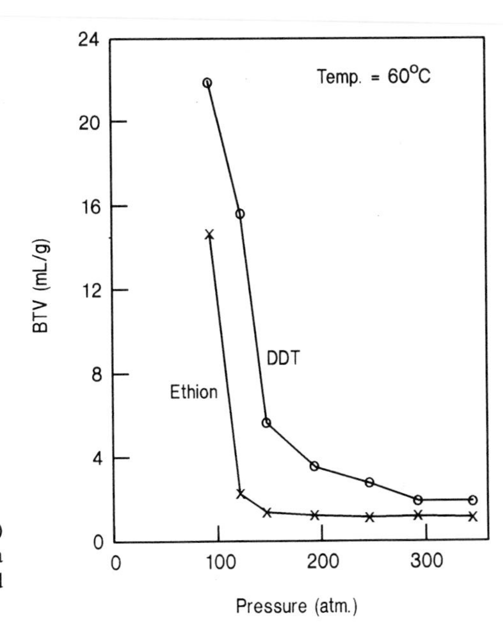

Fig. 6. Breakthrough volumes (BTV) for chlorinated pesticides on alumina as a function of pressure as determined by elution pulse chromatography

it is important to remember that a supercritical fluid will undergo close to a thousand-fold expansion upon decompression onto a "accumulator" cartridge, requiring that a adequate sorbent be available to capture the targent analyte without breakthrough occurring from the trapping cartridge. Similar considerations also apply when collecting or fractionating analytes on sorbents under elevated pressure conditions. As shown in Fig. 6, the breakthrough volume for two pesticides, DDT and ethion, on the sorbent, alumina, decreases with increasing pressure of CO_2. Therefore, capture of two pesticides is best undertaken at pressures below 100 atmospheres, while recovery of the analytes can be rapidly accomplished at pressures above 200 atmospheres. It should be noted that selective desorption of specific analyte classes from particular sorbents can be affected by changing the nature of supercritical fluid as reported by Levy et al. [83] and by Alexandrou and Pawliszyn [9].

3.6 Sample Matrix Effects in SFE

The nature of the sample matrix can have a profound effect on the results that are obtained with SFE. Unfortunately, a knowledge of analyte solubilities in supercritical fluids does not always allow a prediction to be made as to the effectiveness of SFE for extracting a particular matrix [84]. Extraction of real sample matrices, such as soils or biological tissue, should be carried out experimentally, rather then depending on theory or results obtained on neat analytes. In this section we shall examine the factors in the sample matrix which influence the results obtained via SFE.

3.6.1 Physical Matrix Effects

The physical morphology of the substrate undergoing SFE can have a pronounced influence on the efficiency of the extraction and the rate at which it is conducted. In general, the smaller the particle size of the substrate, the more rapid and complete the extraction will be. This effect is largely due to the shorter internal diffusional path lengths over which the extracted solutes must travel to reach the bulk fluid phase. Studies [85] have shown that the geometric size of the matrix particles can influence the speed and completeness with which a SFE can be conducted. As in solid-liquid extraction, an increase in a matrix's porosity will generally promote a more efficient and rapid SFE.

The leaching of a large amount of solute(s) from a sample matrix can weaken the internal structure of the substrate, leading to comminution of the matrix within the extraction cell. A deleterious artifact of this process can be potential plugging of the sample matrix in the extractor. This condition can be partly alleviated by reversing the flow of the extraction gas via a tandem arrangement of valves or by reducing the flow rate of fluid through the matrix.

3.6.2 Chemical Changes in the Sample Matrix

The chemical composition of the sample matrix can have either an enhancing or retarding effect on the results that are obtained with SFE. One of the major parameters that influences the composition of the supercritical fluid extract is the moisture level in the sample matrix. For example, aroma oils from tobacco products are preferentially removed in the absence of moisture, while the presence of water is required for the extraction of alkaloids from the tobacco matrix [86].

The effect of moisture on the SFE of analytes from biological tissues has been a point of controversy for some time among researchers. However, it appears that partial dehydration of the sample matrix will allow a more rapid SFE to be performed. This is due to the fact that highly hydrophilic matrices inhibit contact between the supercritical fluid and the target analytes.

King and coworkers [26] have demonstrated that the removal of water can have a dramatic effect on the recovery of lipid moieties from meat products. Similar trends have been noted by other research groups concerned with the removal of pigments from krill [87] and drugs from body organs

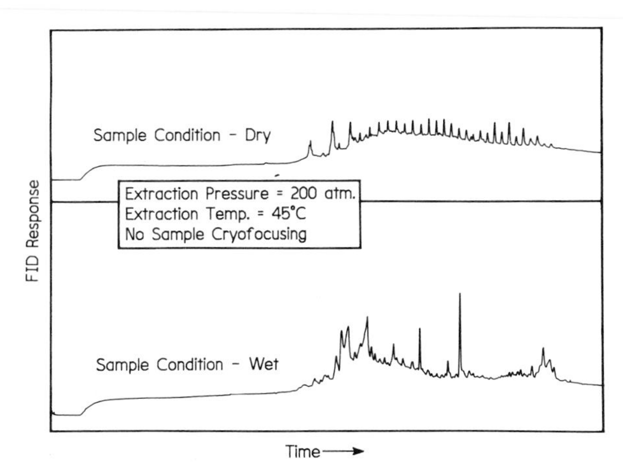

Fig. 7. SFE-SFC profiles of an aquifer solids sample before and after freeze drying of the sample

[62]. However, in some specific cases, the presence of water may actually aid in the recovery of the target analyte by acting as a "internal cosolvent". For this reason, some analysts actually spike the sample matrix with water before performing SFE. A rather dramatic example of the effect of water in the sample matrix on the results of a SFE are shown in Fig. 7. Inspection of the resultant chromatograms shows that the two profiles are different with respect to the number of components and their relative distribution in the sample. These observed differences suggest that the presence of water in the soil matrix may promote the extraction of some components into the fluid phase relative to those obtained from a dry sample.

3.6.3 Impact of Matrix on Extraction Kinetics

The rate of removal of a solute from a matrix using a SFE is a function of its solubility in the fluid media and the rate of mass transport of the solute out of the sample matrix. Rate limiting kinetics can adversely impact on the rapid extraction of an analyte despite favorable its solubility characteristics in the supercritical fluid medium. This situation often occurs when the analyst is trying to isolate a target analyte from a sample matrix, such as a sorbent or a soil.

It is useful to invoke a simple extraction model for the SFE of an analyte from a single particle to visualize the rate inhibiting mechanisms which impact on the extraction. As shown in Fig. 8, there are four major mass transport mechanisms to consider:

- analyte diffusion through the internal volume of the sample
- surface desorption of the analyte

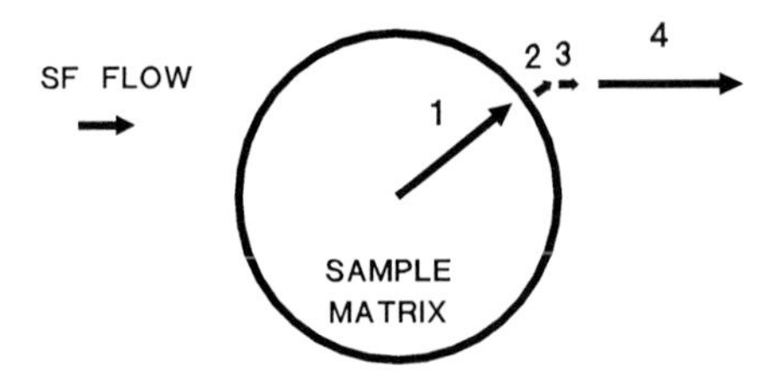

1 = Diffusion through matrix
2 = Desorption from surface
3 = Diffusion through 'SF' surface film
4 = Transport in SF flow

Fig. 8. Mass transport steps for the SFE of an analyte from a porous matrix particle

- diffusion of the analyte through a surface boundary layer
- transport in the bulk supercritical fluid phase

If the rate determining step (RDS) is intraparticle diffusion, then the rate of extraction will be a function of the particle size of the sample matrix. It should be recognized that some matrices when exposed to supercritical fluids swell, thereby facilitating the mass transport of the analyte from within a sample matrix [88]. An excellent example of this principle is the observation that polymeric films are plasticized by supercritical gases [89]. This undoubtably contributes to the recent success achieved by analytical chemists in applying SFE for the analysis of additives in plastics [90].

Surface desorption of an analyte by a supercritical fluid is an important step in SFE for many sample types. For certain analyte-matrix combinations, the "solvent power" of the supercritical fluid alone will not suffice to assure a rapid or complete extraction. Studies on the regeneration of adsorbents [91, 92] have shown that many compounds are note completely recovered with neat supercritical fluids and that desorption times are prohibitively long. The use of a cosolvent, such as water or methanol, will frequently accelerate the desorption of an analyte from the surface of the sample matrix. Wheeler and McNally [15] have shown that extraction efficiencies of herbicides from soils can be increased by direct addition of microliter quantities of ethanol or methanol to the sample before commencement of extraction.

Diffusion of the analyte through a surface boundary layer may also kinetically influence analyte extraction. As noted by King [93] and Parcher [69], many solid samples will promote condensation of a surface layer of the dense extraction fluid at the fluid-solid interface. The density of the adsorbed surface film will partly depend on the pressure applied to the supercritical fluid and the affinity of the sample matrix for the fluid. The development of a condensed fluid film at the surface of the sample matrix can aid in the recovery of certain analytes through competitive adsorption at the sample interface [93] as well as inhibit the transport of the analyte into the fluid phase. The kinetics of transport through a rate limiting surface film will primarily depend on the thickness of the surface film and the total surface area of the sample matrix.

The final step depicted in Fig. 8 is the transport of the analyte in the bulk fluid phase. Such transport is governed primarily by the diffusional coefficient of the analyte in the fluid medium. As noted previously, the diffusion coefficients of solutes in supercritical fluids are intermediate between those that they exhibit in liquid or gaseous media. This factor is independent of the sample matrix. In some instances, enhancement of the mass transfer of an analyte may be expected from free convection effects due to the variable extraction density of the supercritical fluid [94]. Such an effect is readily observed when conducting an extraction in a vertically-orientated cell, where an analyte concentration gradient exists due to the differences in the respective densities of extraction fluid and target analyte.

3.7 Problems and Future Research Needs in Analytical SFE

The successes of analytical SFE have been noted in the previous sections and amply demonstrated in the literature. However, as with any evolving technique, analytical SFE has capabilities and limitations that are not totally understood. Compared to liquid extractions, SFE has several more experimental factors that must be controlled and understood to achieve reproducible results. Some of these parameters are tabulated in Table 3 and have been discussed in previous sections.

Factors such as the collection technique and sample size are determined by the physical nature of the expected extract and sample homogeneity, respectively. Obviously, an extract consisting of volatiles cannot be collected by employing a simple phase separation. In this case, a packed accumulator cartridge would be the preferred collection device. Small samples should also be avoided in cases where the sample matrix is not homogeneous, since

Table 3. Experimental factors affecting analytical SFE

Pressure
Temperature
Flow rate
Extraction time
Collection technique
Sample size
Choice of supercritical fluid
Choice of modifier
Amount of modifier
System leaks
System contamination
Sample matrix

Table 4. Reproducibility of analytical scale supercritical fluid extractions

Analyte	Matrix	Concentration range (ppm)	% RSD[a]
Polyaromatic Hydrocarbons (PAH)	Diesel exhaust Particulate (NIST)	1.4 – 55	0.3 – 7.0
PAH	Urban dust (NIST)	2.0 – 8.2	0.5 – 1.0
PAH	Urban dust (NIST)	2.1 – 8.1	0.4 – 1.0
PAH	Tenax	0.1 – 2.0	3
PAH	River sediment	–	2 – 20
PAH	Lampblack	–	2.4 – 16
Dioxin	Soil	0.0016 – 0.0082	1 – 20
Spice components	Basil	–	6 – 17
Polymer additives	Polyethylene	5.5 – 1300	4 – 29
Menadione	Rodent feed	20 – 1500	0.4 – 4.7

[a] Percent relative standard deviation.

the resultant extract may not accurately reflect the content of the sample. Utilization of larger sample sizes in analytical SFE obviously favors off-line techniques, since on-line methodology is ultimately limited by the analyte concentrations that can be chromatographed.

The set of parameters listed in Table 3 makes optimization of analytical SFE to a particular analysis problem more time consuming than conventional extraction techniques. However, the application of statistical experimental design methods to SFE [95] promises to ease the burden on the analytical chemist. Excellent extraction reproducibility has been reported for a number of different sample types using SFE as shown in Table 4.

Coextraction of unwanted solutes along with the target analyte frequently occurs in analytical SFE, whether conducted in the off- or on-line mode. These interferences can be removed either by conventional sample cleanup methods or by utilizing a sorbent column downstream from the SFE device. Selective desorption of target analytes can be affected from coextracted background matrix components, provided there are sufficient differences in their respective breakthrough volumes on the sorbent in the presence of the supercritical fluid (see Fig. 6). Alternatively, one can use a selective detector which "blanks out" the interfering species in the analysis step. An example of this principle is shown in Fig. 9 where an on-line extraction has been performed on the pesticide, DDT, both neat and in a fat matrix with potentially interfering components. The use of an electron capture detector (ECD) after SFE and capillary SFC shows high specificity for the chlorinated pesticide, whereas a flame ionization detector (FID) trace on the same extract shows only a large unresolved triglyceride profile.

Finally, the opportunities for applying SFE in analytical chemistry are numerous and appear to be only partially offset by the above limitations. Routine use of the technique will require the development of instrumentation that will allow the analysis of multiple samples in either a batch or

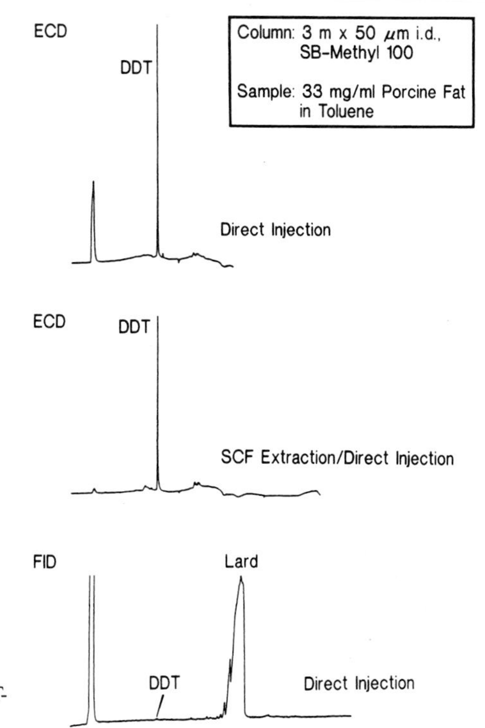

Fig. 9. Detector specificity for DDT-spiked pork fat by SFC and SFE-SFC

serial mode. These and other developments portend a promising future for SFE: a technique which will help the analytical chemist and improve his laboratory environment.

References

1. Majors RE (1989) LC-GC 7:92
2. Steinwandter H (1989) In: Zweig G, Sherma J (eds) Analytical methods for pesticides and plant growth regulators, Vol 17. Academic, New York, pp 63–71
3. Giddings JC, Myers MN, King JW (1969) J Chromatogr Sci 7:276
4. Sie ST, Rijnders GWA (1967) Separation Sci 2:755
5. Stahl E, Schilz W, Schutz W, Willing E (1978) Angew Chem Int Ed Engl 17:731
6. Stahl E, Quirin KW, Gerard D (1988) Dense gases for extraction and refining. Springer-Verlag, Berlin
7. Hawthorne SB, Miller DJ (1986) J Chromatog Sci 24:258
8. Schantz MM, Chesler SN (1986) J Chromatog 363:397
9. Alexandrou N, Pawliszyn J (1989) Anal Chem 61:2770
10. Hawthorne SB, Krieger MS, Miller DJ (1989) Anal Chem 61:736
11. Onuska FI, Terry KA (1989) J High Resolut Chromatog 12:357
12. Dooley KM, Kao CP, Gambrell RP, Knopf FC (1987) Ind Eng Chem Res 26:2058

13. King JW (1989) J Chromatog Sci 27:355
14. Nam KS, Kapila S, Pieczonka G, Clevenger TE, Yanders AF, Viswanath DS, Mallu B (1988) In: Perrut M (ed) Proceedings of the international symposium on supercritical fluids. Tome 2 Institute National Polytechnique de Lorraine, France, p 743–750
15. Wheeler JR, McNally ME (1989) J Chromatog Sci 27:534
16. Janda V, Steenbeke G, Sandra P (1989) J Chromatog 479:200
17. Lopez-Avila V, Becker WF, Billets S (1989) In: Proceedings of the 5th annual waste testing and quality assurance. Symposium, Vol 2. US EPA, Washington, DC, p 11–73
18. Schafer K, Baumann W (1989) Z Anal Chem 332:884
19. Schneiderman MA, Sharma AK, Locke DC (1987) J Chromatog 409:343
20. Schneiderman MA, Sharma AK, Locke DC (1988) J Chromatog Sci 26:458
21. Xie QL, Markides KE, Lee ML (1989) J Chromatog Sci 27:365
22. Hawthorne SB, Miller DJ (1987) J Chromatog 403:63
23. McHugh MA, Krukonis VJ (1986) Supercritical fluid extraction: principles and practice. Butterworths, Boston, MA
24. Zosel K (1980) In: Extraction with supercritical gases. Verlag Chemie, Weinheim, Germany, pp 1–23
25. Eggers R, Tschiersch R (1978) Chem Ing Tech 50:842
26. King JW, Johnson JH, Friedrich JP (1989) J Agric Food Chem 37:951
27. Giddings JC, Myers MN, McLaren L, Keller RA (1968) Science 162:67
28. King JW (1984) Preprints-Polymeric Materials Science & Engineering Division, Vol 51. American Chemical Society, Washington, DC, pp 707–712
29. Stahl E, Willing E (1980) Pharm Ind 42:1136
30. Hemmaplardh B, King Jr AD (1972) J Phys Chem 76:2170
31. Brogle H (1982) Chem Ind (12):385
32. Francis AW (1954) J Phys Chem 58:1099
33. Lira CT (1988) In: Charpentier BA, Sevenants MR (eds) Supercritical fluid extraction and chromatography. American Chemical Society, Washington DC, pp 1–25
34. Chimowitz EH, Pennisi KJ (1986) AIChE J 32:1665
35. Johnston KP, Barry SE, Read NK, Holcomb TR (1987) Ind Eng Chem Res 26:2372
36. Peter S, Brunner G, Riha R (1976) Fette Seifen Anstrichmittel 76:45
37. Dobbs JM, Wong JM, Lahiere RJ, Johnston KP (1987) Ind Eng Chem Res 26:56
38. Wong JM, Johnston KP (1986) Biotech Progress 2:29
39. Dobbs JM, Johnston KP (1987) Ind Eng Chem Res 26:1476
40. Hedrick J, Taylor LT (1989) Anal Chem 61:1986
41. Johnston KP (1989) In: Johnston KP, Penninger JML (eds) Supercritical fluid science and technology. American Chemical Society, Washington, DC, pp 1–12
42. Rizvi SSH, Benado AL, Zollweg JA, Daniels JA (1986) Food Tech 40 (6):55
43. Friedrich JP, List GR, Spencer GF (1988) In: Baldwin AR (ed) Proceedings of the 7th international conference on jojoba and its uses. American Oil Chemists' Society, Champaign, IL, 165–172
44. Hawthorne SB, Miller DJ, Krieger MS (1988) Fresenius Z Anal Chem 330:211
45. King JW (1984) J Am Oil Chemists' Soc 61:689
46. List GR, Friedrich JP, King JW (1989) Oil Mill Gazetteer 95 (6):28–34
47. Bowman Jr LM (1976) Dense gas chromatographic studies. Ph D Thesis, University of Utah, Salt Lake City, UT
48. Prausnitz JM (1969) Molecular thermodynamics of fluid-phase equilibria. Prentice-Hall, Englewood Cliffs, NJ, p 36–39
49. Kurnik RT, Reid RC (1981) AIChE J 27:861
50. Allada SR (1984) Ind Eng Chem Process Des Dev 23:344
51. Fedors RF (1974) Polym Eng Sci 14:147
52. Jayasri A, Yaseen M (1980) J Coatings Tech 52 (667):41–45
53. Flory PJ (1953) Principles of polymer chemistry. Cornell University Press, Ithaca, NY
54. Sugiyama K, Saito M, Hondo T, Senda M (1985) J Chromatogr 332:107
55. Jackson WP, Markides KE, Lee ML (1986) HRC & CC 9:213
56. Skelton Jr RJ, Johnson CC, Taylor LT (1986) Chromatographia 21:4

57. Gmuer W, Bosset JO, Plattner E (1987) J Chromatogr 388:335
58. Engelhardt H, Gross A (1988) HRC & CC 11:38
59. Raynor MW, Davies IL, Bartle KD, Clifford AA, Williams A, Chalmers JM, Cook BW (1988) HRC & CC 11:766
60. McNally ME, Wheeler JR (1988) J Chromatog 435:63
61. Saito M, Yamaguchi Y, Inomata K, Kottkamp W (1989) J Chromatog Sci 27:79
62. Ramsey ED, Perkins JR, Games DE, Startin JR (1989) J Chromatogr 464:353
63. Thiebault D, Chervet JP, Vannoort RW, DeJong GJ, Brinkman UA, Frei RW (1989) J Chromatogr 477:151
64. Sieber R (1988) In: Perrut M (ed) Proceedings of the international symposium on supercritical fluids, Vol 2. Institute National Polytechnique de Lorraine, France, p 619
65. van Wasen U, Swaid I, Schneider GM (1980) Angew Chem Int Ed Engl 19:575
66. Smith RD, Udseth HR, Wright BW, Yonker CR (1987) Sep Sci Tech 22:1065
67. Feist R, Schneider GM (1982) Sep Sci 17:261
68. Sie ST, Van Beersum W, Rijnders GWA (1966) Sep Sci 1:459
69. Parcher J, Strubinger JR (1989) J Chromatogr 479:251
70. King JW, Eissler RL, Friedrich JP (1988) In: Charpentier BA, Sevenants MR (eds) Supercritical fluid extraction and chromatography. American Chemical Society, Washington, DC, p 63–88
71. Markides KE, Lee ML (1989) SFC applications. Brigham Young University Press, Provo, UT
72. Smith RD, Udseth HR, Wright BW (1985) In: Penninger JML, Radosz M, McHugh MA, Krukonis VJ (eds) Supercritical fluid technology. Elsevier, Amsterdam, pp 191–223
73. McMahon D, Cohen KA, Taylor LT (eds) (1988) Chromatography illustrated. Preston Publications, Niles, IL, pp 73–90
74. Fields SM, Grolimund K (1988) HRC & CC 11:727
75. McNally ME, Wheeler JR (1988) J Chromatogr 447:53
76. Klesper E, Leyendecker D, Schmitz FP (1986) J Chromatogr 366:235
77. Levy JM, Ritchey WM (1986) J Chromatogr Sci 24:242
78. Hawthorne SB, Miller DJ (1987) Anal Chem 59:1705
79. Engelhart WG, Gargus AG (1988) 20 (2):30
80. Swanson JT, Andersen MR, Richter BE (1990) Theory of mass transfer applied to micro-SFE. Paper presented at the symposium/workshop on supercritical fluid chromatography, Part City, UT
81. Hawthorne SB (1990) Anal Chem 62:633A
82. Andersen MR, Swanson JT, Porter NL, Richter BE (1989) J Chromatogr Sci 27:371
83. Levy JM, Cavalier RA, Bosch TN, Rynaski AF, Huhak WE (1989) J Chromatogr Sci 27:341
84. King JW, France JE, Taylor SL (1990) In: Sandra P, Redant G (eds) Proceedings – 11th international symposium on capillary chromatography. Hüthig, Heidelberg, pp 595–601
85. Synder JM, Friedrich JP, Christianson DD (1984) J Am Oil Chemists' Soc 61:5475
86. Roselius W, Vitzhum OG, Hubert P (1976) DBP 2142205
87. Yamaguchi K, Murakami M, Nakano H, Konosu S, Kokura T, Yamamoto H, Kosaka M, Hata K (1986) J Agric Food Chem 34:904
88. King JW (1989) Abstracts of the Pittsburgh conference, Atlanta, GA, #397
89. Chiou JS, Barlow JW, Paul DR (1985) J Apply Polym Sci 30:3911
90. Nicholas J, Bartle KD, Clifford AA (1990) In: Sandra P, Redant G (eds) Proceedings – 11th international symposium on capillary chromatography. Hüthig, Heidelberg, pp 665–672
91. Modell M, deFilippi RP, Krukonis V (1980) In: Suffet IH, McGuire MJ (eds) Activated carbon adsorption of organics from the aqueous phase, Vol 1. Ann Arbor Science Publishers Inc, Ann Arbor, MI, pp 447–462

92. Picht RD, Dillman TR, Burke DJ, deFilippi RP (1982) In: Ma YH (ed) Recent advances in adsorption and ion exchange, Vol 78. AIChE, New York, NY, pp 136–149
93. King JW (1987) In: Squires TG, Paulaitis ME (eds) Supercritical fluids: Chemical and engineering principles and applications. American Chemical Society, Washington, DC, pp 150–171
94. Randolph TW (1990) Trends in Biotech 8:78
95. Fisher RJ (1989) Food Tech 43(3):90

4 Coupled Supercritical Fluid Extraction-Capillary Gas Chromatography (SFE-GC)

STEVEN B. HAWTHORNE

4.1 Introduction

A frequently held misconception is that extracts resulting from supercritical fluid extraction (SFE) must be analyzed using supercritical fluid chromatography (SFC). While SFC is certainly useful for separating many analytes from SFE extracts, SFE is an independent sample preparation technique, and extracts can be analyzed using a wide variety of techniques including (but not limited to) chromatographic, spectroscopic, radiochemical, electrochemical, and gravimetric. When analytes have sufficient vapor pressures, capillary gas chromatography is the method of choice because of its very high resolution per unit time, and the excellent variety of detectors that are available. Since the most commonly used supercritical fluids (e.g., CO_2) are gases at ambient conditions, SFE can also be directly coupled with capillary GC (SFE-GC) using simple techniques.

The direct coupling of SFE with GC has two advantages over performing SFE off-line. First, SFE-GC eliminates all sample handling (and associated errors) between extraction and chromatographic analysis. Second, SFE-GC can yield maximum sensitivity since (with certain coupling techniques), 100% of every extracted species can be transferred into the GC column, and thus to the detector. This chapter will discuss experimental techniques, special abilities and limitations, quantitative aspects, and applications of SFE coupled with capillary GC.

4.2 Performing SFE-GC

4.2.1 Instrumentation and Methods

Successful performance of SFE-GC requires three basic steps: 1) extraction of the target analytes into the supercritical fluid, 2) depressurization of the supercritical fluid to ambient pressure and coincident trapping and focusing of the extracted analytes, and 3) GC separation of the target analytes. Since the analytes are extracted over a period of time (typically 5 to 30 minutes) that is long compared to a GC peak width, they must be refocused prior to GC separation to obtain good chromatographic peak shapes. The most important variations in the approaches used for SFE-GC occur in the methods used for collecting and focusing the extracted analytes at the head of the chromatographic system during the extraction step. The essential problem to be solved for quantitatively trapping and focusing the analytes from the SFE effluent is based on the high gas flows encountered upon depressurization. For example, extraction with 1 ml/min of supercritical CO_2 results, upon depressurization, in a CO_2 gas flow of ca. 500 ml/min. Thus, the trapping method must be capable of quantitatively collecting the analytes under relatively high gas flow conditions.

Two general approaches that have been used are: 1) collecting the analytes in an accumulating device external to the GC [1–3], and 2) using the GC column itself (or a retention gap) for analyte collection [4–12]. For external collections, the analytes are depressurized into a cold trap or sorbent resin [1–3]. After the SFE is completed, the trap is heated and the analytes are swept into the GC by the carrier gas. The use of traps placed external to the column is attractive since the SFE fluid does not have to be vented into the GC, and any potential effect of the fluid on the column or detector can be ignored. Unfortunately, this approach has been less successful in producing quantitative results than approaches that use the GC column for collection of the extracted analytes. Since analytes that are amenable to GC analysis have significant vapor pressures, empty cold traps may have little utility for SFE-GC (especially when the high gas flows are considered), and quantitative collection of extracted analytes has not been reported for this technique [1–3]. However, the use of sorbent traps (e.g., Tenax) that can be thermally desorbed to recover the analytes may be more promising and warrants further development.

The majority of published SFE-GC analyses have utilized the cooled GC column to collect and focus the extracted analytes from the depressurized supercritical fluid, and this approach will be the focus of subsequent discussions. SFE-GC coupling is performed by inserting the extraction cell outlet restrictor directly into a conventional GC split injection port (split SFE-GC) [5–9], or by inserting the restrictor into the capillary GC column through a conventional on-column injection port (on-column SFE-GC) [10–12]. Neither of these approaches requires significant modification of conventional GC instrumentation, and the same GC can be used for conventional liquid solvent injections with little or no conversion.

Split and on-column SFE-GC approaches are shown schematically in Fig. 1. With split SFE-GC, the extraction cell outlet restrictor (typically a length of 15- to 30-μm i.d. fused silica tubing) is inserted through the injection port septum [5–7] or a septumless injector [8] into the injection port liner. Both arrangements result in a gas-tight system which allows the split ratio to be controlled while SFE is conducted. During the extraction the SFE effluent (arrows) is depressurized inside of the split injection port liner and, analogous to a conventional split injection, a fraction of the extracted analytes (small circles) enters the GC column for cryogenic focusing in the GC column stationary phase, while the remainder is flushed out the split vent. The steps for split SFE-GC are: 1) cool the oven to the cryogenic trap-

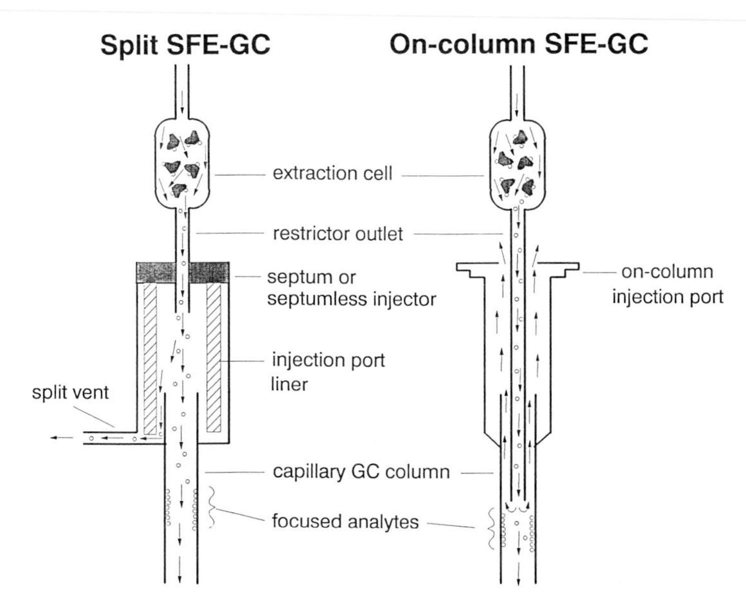

Fig. 1. Schematic diagrams of split and on-column SFE-GC. The extraction fluid is represented by *arrows* and the analytes are represented by *small circles*. A discussion of the two coupling techniques is given in the text

ping temperature, 2) insert the restrictor into the injection port, 3) perform the extraction (typically 10 minutes), 4) remove the restrictor from the injection port, and sweep the CO_2 (or other extraction fluid) from the GC column with the GC carrier gas, and 5) perform a normal temperature programmed GC analysis. With split SFE-GC, it is helpful to block the GC carrier gas flow during the SFE step with a shut-off valve placed near the injection port to avoid having the SFE effluent back up into the carrier gas lines. The split ratio can then be conveniently controlled with a needle valve placed on the split outlet line [8].

On-column SFE-GC is performed by inserting the fused silica extraction cell restrictor (150 μm o.d.) into the capillary GC column (250 or 320 μm i.d.) through a conventional on-column injector [10–12], and the steps used for analysis are essentially identical to those listed above for split SFE-GC. In contrast to the gas-tight system used for split SFE-GC, where both the depressurized supercritical fluid and the extracted analytes are divided between the capillary column and the split flow, all of the analytes (small circles) extracted using on-column SFE-GC are deposited directly in the GC column stationary phase. The gaseous CO_2 (or other extraction fluid represented by arrows in Fig. 1) is continuously vented back through the injection port, and to a lesser extent, through the GC column. (For additional experimental details on performing split SFE-GC and on-column SFE-GC, the reader should consult references 5–12.)

4.2.2 Fluids and Extraction Conditions Used for SFE-GC

As is the case for most SFE studies, CO_2 has been most popular for SFE-GC, although other fluids that are gases at ambient pressure including N_2O and SF_6 have also been used. While CO_2 has several practical advantages for SFE-GC (e.g., high purity, low cost, low toxicity and reactivity) and is a good first choice for performing SFE-GC analyses, other fluids should be tested when the recoveries obtained using CO_2 are not quantitative. The choice of a supercritical fluid (and extraction pressure) based on analyte solubility alone is not always successful, since solubility considerations do not include the ability of the supercritical fluid to displace the analyte from the active sites on the sample matrix. For example, CO_2 and N_2O have very similar solvent strengths [13], but N_2O has been shown to recover PAHs from some matrices much faster than CO_2 [8]. Fortunately, species that can be analyzed using capillary GC are relatively non-polar, and as a general rule, can be quantitatively extracted with either CO_2 or N_2O in reasonable times (e.g., 10 to 20 minutes).

In most cases, SFE-GC extractions are performed at 300 to 400 atm, pressures that yield relatively high solvent strengths for CO_2 and N_2O, and are relatively easy to achieve within the constraints of commercial pump and extraction cell pressure capabilities. However, low extraction pressures can be used to selectively extract very non-polar organics. For example, SFE-GC/MS analysis of urban air particulates extracted with 75 atm of N_2O recovered 90 to 95% of the C22 to C26 *n*-alkanes, while polycyclic aromatic hydrocarbons (PAHs) such as benzo[a]pyrene and benzo[ghi]perylene were not extracted. The second SFE-GC/MS analysis of the same sample using a 300-atm extraction quantitatively recovered the PAHs [14]. Extraction temperatures are typically a few degrees above the critical temperature of the fluid, i.e., 40 to 60 °C.

While the choice of supercritical fluid for SFE-GC depends mostly on its ability to quantitatively extract the target analytes, the potential effect of the fluid on the GC detector (unless trapping external to the GC is utilized) should also be considered. For example, using SF_6 for SFE-GC with an electron capture detector would result in excessively high baselines. The use of CO_2 presents few problems for most GC detectors (with the possible exception of FTIR), and our experience demonstrates that, after an extraction is completed, both CO_2 and N_2O are completely flushed from the GC column by the carrier gas almost immediately, and thus do not affect detector response during the chromatographic separation of the extracted analytes. When SFE-GC/MS is performed using a wide-bore (320 μm i.d.) column, it is possible to generate excessively high pressures (e.g., 10^{-4} torr) in the MS ion source during the extraction step. However, this can be easily avoided by either using a 250-μm i.d. column, or by placing a 1-m section of 150-μm i.d. transfer line between the 320-μm i.d. column and the ion source [10, 11]. Some safety concerns exist with N_2O since it is an oxidant, and it may be best to avoid its use for the extraction of large samples with high organic content, and with the use of hydrogen carrier gas for SFE-GC.

4.3 Abilities and Limitations of Split and On-Column SFE-GC

A comparison of split and on-column SFE-GC shares many characteristics with a comparison of conventional GC split and on-column injections of liquid solvent extracts. Practical comparisons of present versions of these two techniques are listed in Table 1 and are discussed below. It is important to note, however, that these comparisons should only be used as a frame of reference for SFE-GC methods developed so far, and will certainly change as SFE-GC methods continue to mature.

4.3.1 Chromatographic Peak Shapes Obtained Using SFE-GC

As discussed above, the quality of the chromatograms obtained using SFE-GC depends upon the ability of the coupling method to collect and focus the extracted analytes during the extraction step. When properly performed, both on-column and split SFE-GC yield chromatographic peak shapes that are virtually identical to those obtained using conventional GC injections, as demonstrated in Fig. 2 by comparisons of conventional split and on-column GC injections of methylene chloride extracts with those obtained using each SFE-GC technique for the analysis of a sediment contaminated with a coal-derived anthracene oil, and terpenoid compounds from freshly picked juniper bush needles. (It is interesting to note that the SFE-GC

Table 1. Comparison of existing on-column and split SFE-GC techniques

Characteristic	On-Column SFE-GC	Split SFE-GC
% of analyte transferred into GC column	100%	0.2 to 5%
Reproducibility (RSD of raw peak areas)	2–10%	2–10%
Most volatile alkane efficiently trapped	C7 to C8	smaller than C6
Maximum flow rate (as supercritical fluid)	0.4 ml/min	2 ml/min
Sample size (10 min quantitative extraction)	1 mg to 1 g	1 mg to 15 g
Can be used with modifiers	no?	yes
Wet samples	difficult	yes
Works with high amounts of extractable matrix species	rarely	often
Restrictor plugs during extraction	some samples	no
Conversion required to conventional GC	none	minor
Typical total extraction and GC analysis time	1 hour	1 hour

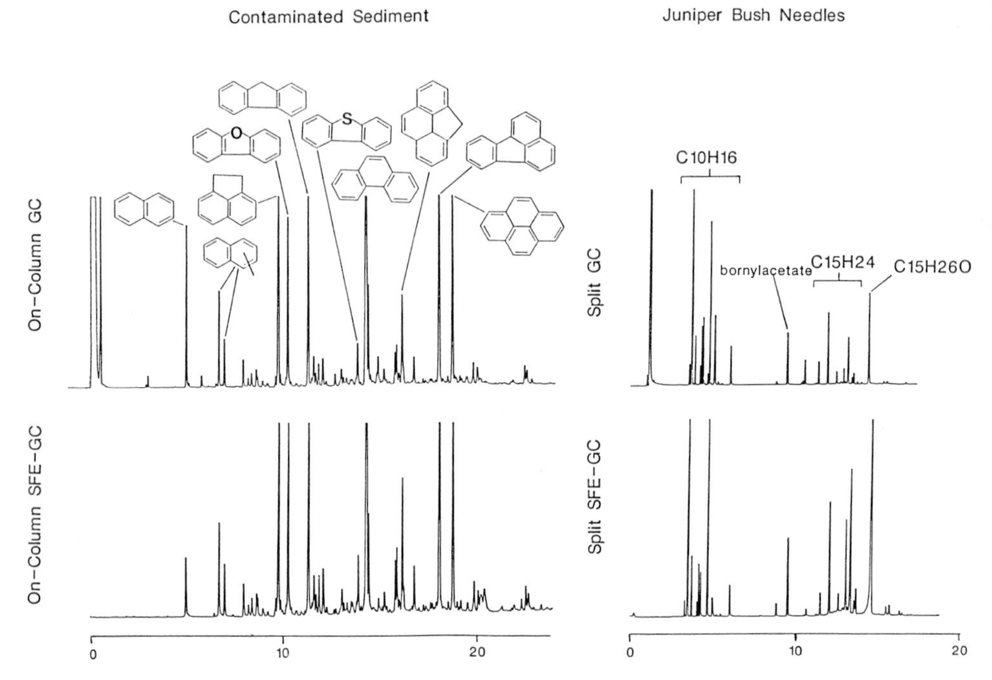

Fig. 2. Comparison of chromatographic peak shapes generated using conventional GC injections of liquid solvent extracts with those obtained using SFE-GC/FID analysis. The juniper bush needles were analyzed using split SFE-GC (split ratio ca. 1 : 50) with a 1-min extraction with 400 atm CO_2 (45 °C) and a supercritical fluid flow rate of ca. 0.5 ml/min. The GC oven was held at −30 °C during the SFE step, then rapidly heated to 70 °C followed by a temperature ramp of 8 °C/minute to 300 °C. The methylene chloride extract was prepared by extracting 10 grams of needles for 24 hours and concentrating to ca. 2 mL. A 1 μL aliquot was injected at 70 °C followed by a temperature ramp of 8° to 300 °C. The sediment contaminated with a coal-derived anthracene oil was analyzed in a similar manner using on-column SFE-GC, and a conventional on-column injection of a concentrated methylene chloride extract (1 μL) as previously described [14]. A 30 m J&W DB-5 (320 μm i.d., 1 μm film thickness) GC column was used for each analysis. Peak identifications for both samples were based on SFE-GC/MS analyses of replicate samples

analysis with a one minute extraction yielded a higher proportion of the sesquiterpenes from the bush needles than the 24 hour methylene chloride extraction.)

The cryogenic trapping temperature used during the extraction obviously affects the ability to efficiently focus volatile analytes, and split SFE-GC can yield good peak shapes for species at least as volatile as hexane by trapping at −50 °C [8]. On-column SFE-GC with CO_2 has been limited to trapping temperatures no lower than ca. −30 to 0 °C (depending on the capillary column used) because colder trapping temperatures combined with the cooling from the expanding CO_2 cause the GC column to become plugged with frozen CO_2. Under these conditions, species more volatile than n-octane show some peak broadening [11]. Both on-column and split SFE can also yield poor chromatographic peak shapes when the extraction is performed using excessive supercritical fluid flow rates. On-column SFE yields good peak shapes with supercritical fluid flow rates up to ca. 0.4 ml/min (ca. 200 ml/min gas flow), while split SFE can accommodate flow rates up to ca. 2 ml/min (ca. 1000 ml/min gas flow) [8, 10].

4.3.2 Quantitation Using SFE-GC

Both split and on-column SFE-GC yield quantitative reproducibilities that compare favorably with those obtained for conventional GC injections of liquid solvent extracts. For example, the relative standard deviations in raw peak areas ranged from 1 to 5% for aromatic hydrocarbons using both split SFE-GC and autosampler injections of a liquid solvent standard containing the same test compounds [7]. Reproducibilities in raw peak areas using on-column SFE-GC of samples such as basil spice were also essentially the same as those from replicate conventional on-column injections of methylene chloride extracts, and typically ranged from 2 to 11% for the individual components [11].

The ability to perform quantitative extraction and GC analyses using both split and on-column SFE-GC has been demonstrated for several sample matrices and analyte combinations ranging from environmental pollutants extracted from sorbent resins and sediments, to flavor and fragrance compounds extracted from food products [7–12, 15]. Spike recovery studies, comparison to conventional liquid solvent extractions, and performance of multiple extractions of a single sample have all been used to demonstrate the quantitative abilities of SFE-GC. Perhaps the most convincing demonstrations have been performed using certified standard reference materials from the United States National Institute of Standards and Technology (NIST). Two materials, urban air particulates (SRM 1649) and marine sediment (SRM 1941) have been extensively characterized by NIST using 16- to 48-hour Soxhlet extractions, and the concentrations of several PAHs have been certified for each material. Figure 3 shows typical results for the use of split and on-column SFE-GC/MS for the quantitation of representative PAHs from the air particulates and the marine sediment. Note that the SFE-GC methods yielded excellent quantitative agreement with the certified values, but decreased the time needed for extraction from 48 hours to 20–30 minutes for the air particulates, and from 16 hours to 10 minutes for the marine sediment.

4.3.3 Speed of SFE-GC Analyses

Because SFE extractions typically require less than 30 minutes to perform, and since no concentration and sample handling steps are required between

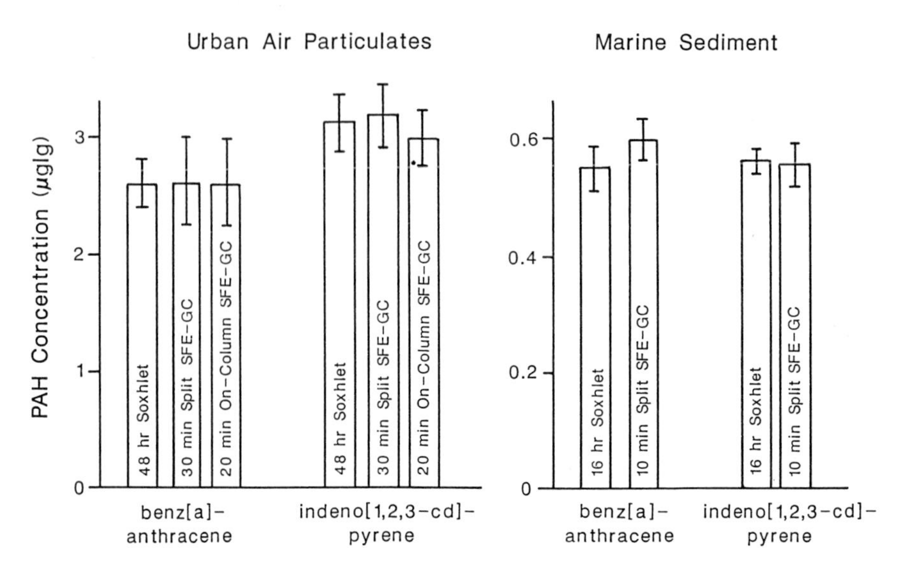

Fig. 3. Quantitation of representative PAHs in standard reference materials, urban air particulates and marine sediment, using conventional liquid solvent (Soxhlet) extractions and split and on-column SFE-GC. The extraction times required for liquid solvent and SFE-GC extractions are shown on each bar. Results are adapted from Refs. [7, 8, and 10]

the SFE and GC steps, SFE-GC techniques are capable of very rapid sample turn-around. For example, each SFE-GC analysis of the standard reference materials described above (Fig. 3) was completed in a total time of less than one hour per sample (compared to days for the conventional methods). The ability of SFE-GC techniques to yield fast turn-around is also demonstrated in Figure 4 by the analysis of roofing tar volatiles collected on a polyurethane foam (PUF) sorbent resin [15]. The total time elapsed between beginning collection of the air sample and completing the GC/MS analysis (excluding data reduction) was one hour and, as shown by the selected ion plots in Fig. 4, ca. one hundred different organic compounds representing five different compound classes were determined.

4.3.4 Sensitivity and Sample Sizes

With conventional liquid solvent injections it is difficult to transfer more than 1% of an extract into a capillary GC column (e.g. a 1-µl on-column injection of a sample concentrated to 100 µL). However, on-column SFE yields maximum sensitivity by transferring every extracted analyte molecule into the GC column, while split SFE-GC transfers ca. 0.2 to 5% of the analytes (depending on the split ratio). On-column SFE-GC is therefore the choice for trace analysis, particularly for samples that are difficult to collect in large quantities (e.g., air-borne particulate matter). Assuming that the

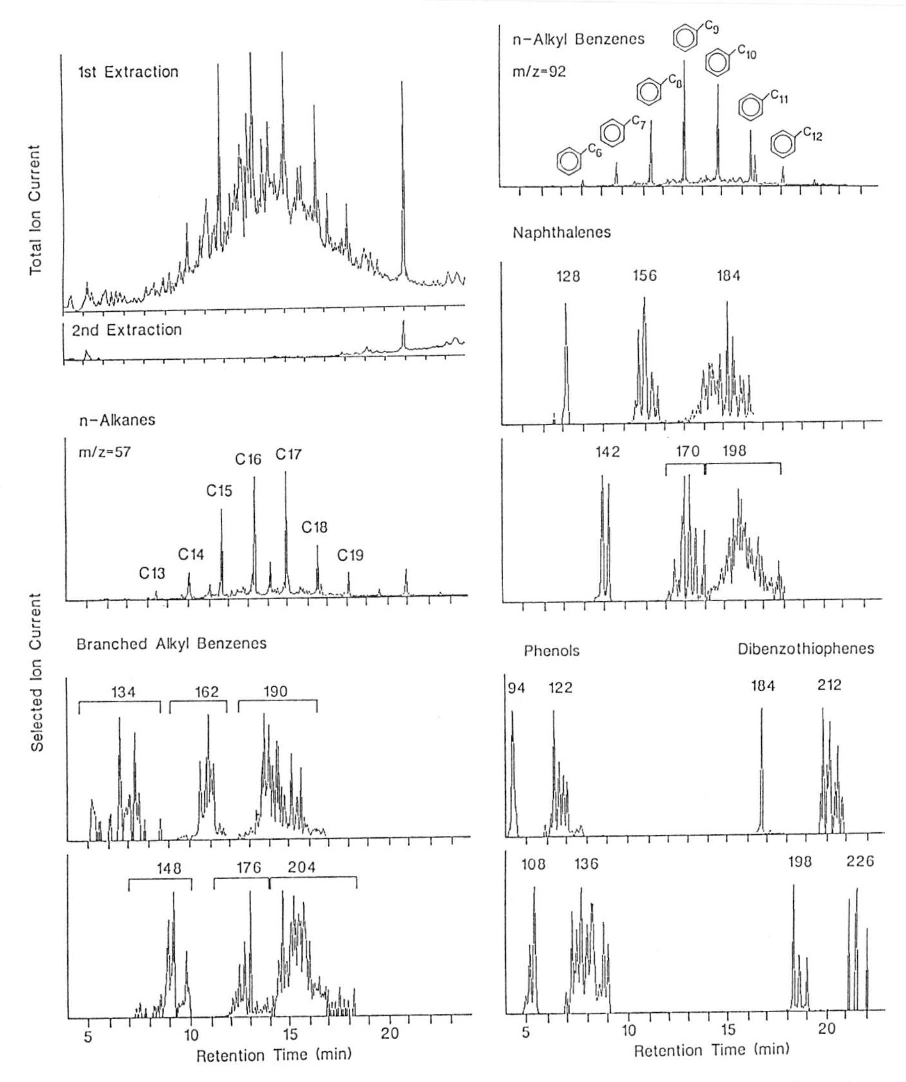

Fig. 4. On-column SFE-GC/MS analysis of roofing tar volatiles collected onto a polyurethane (PUF) sorbent plug near the tar vat operator's face. The numbers above the chromatographic peaks are the masses of the ions used to generate the selected ion plots. Total sample collection, extraction, and GC/MS analysis time (excluding data reduction) was one hour. Adapted with permission from Ref. [15]

GC detector has a detection limit in the pg range for a particular analyte, on-column SFE-GC would have a detection limit of 1 ppb (pg/mg) with a sample as small as 1 mg, or 1 ppt for a 1-gram sample. For the analysis of the standard reference materials shown in Fig. 3, the sample size required was reduced from one to several grams for conventional liquid solvent extraction to only 1–30 mg for SFE-GC.

Because of the sensitivity possible with SFE-GC, the purity of the entire SFE system including the supercritical fluid and sorbent resins (when used), and the elimination of any contaminants introduced from the extraction cells and associated valves must be rigorously controlled, particularly for on-column SFE-GC. The analyst must also be constantly aware that every impurity in the system that is extracted by (or present in) the supercritical fluid will be transferred to the GC and focused in the same manner as the target analytes. Thus, contaminants that are insignificant for SFC may cause considerable artifact peaks in SFE-GC. In order to avoid such artifacts, the purest available supercritical fluids should be used, and the number of potential contaminating components (e.g., valve packings) should be minimized.

While the sizes of samples required to obtain a particular detection limit can be dramatically reduced using SFE-GC, the limitations on the flow rates of the supercritical fluid also limit the maximum sample size that can be quantitatively extracted using SFE-GC. Since many additional factors, including the kinetics and mechanisms controlling the SFE step and the void space in the sample cell also control extraction rates, maximum sample sizes for split and on-column SFE will vary depending on the sample characteristics and the time the analyst is willing to spend on the extraction. Assuming that an extraction should be quantitative in 30 minutes or less, present techniques for on-column and split SFE-GC have been useful for samples as large as ca. 1 gram and 15 grams, respectively [8]. Further improvements in trapping techniques and understanding of extraction kinetics and mechanisms should make SFE-GC useful for larger samples.

4.3.5 Sample Types and Matrix Considerations

SFE-GC has been applied to a wide variety of samples ranging from organic pollutants collected on sorbent resins and found in soils, sediments, and air particulates, to flavor and fragrance compounds from food products. The applications of SFE-GC have recently been reviewed [16], but only represent a fraction of the samples that SFE-GC should be useful for. However, some samples have matrices which make their analysis by SFE-GC impossible with present techniques. Samples containing high concentrations of components that are easily extracted, but not volatile enough for GC analysis (e.g., fats from cheese or meat products) are unsuitable for direct analysis using either on-column or split SFE-GC.

Samples containing high concentrations of water are often difficult to analyze using on-column SFE-GC because the water tends to freeze at the end of the outlet restrictor, thus stopping the flow of the supercritical fluid. However, split SFE-GC works well, even with very wet samples, since the depressurization of the supercritical fluid occurs in the heated injection port, thus preventing ice formation. Wet samples are easily analyzed using split SFE-GC by simply maintaining the GC cryogenic trapping tempera-

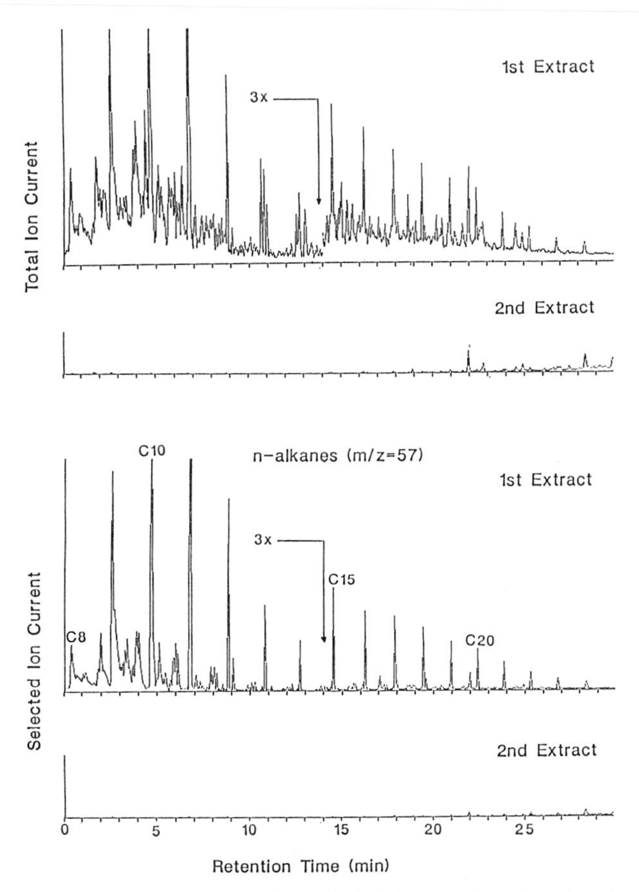

Fig. 5. Split SFE-GC/MS analysis of a wet (20% water by weight) fuel-contaminated sediment. The top two total ion current chromatograms show the first and second SFE-GC/MS analyses of the sediment using 10-minute extractions with 380 atm CO_2 at 50 °C. The bottom two selected ion current chromatograms (m/z = 57) show the alkanes for the first and second SFE-GC/MS analyses. (The relative sensitivity of each chromatogram was multiplied by 3 at a retention time of 14 minutes.) Numbers above the peaks indicate the *n*-alkane chain length. Adapted with permission from Ref. [8]

ture above 0 °C, as demonstrated in Fig. 5 by the SFE-GC/MS analysis of a wet sediment (20% water by weight) that was collected from an aquifer which had been contaminated by leaks from a gasoline refinery [8]. Split SFE-GC has also been successfully used for extractions that require polarity modifiers [7].

The heat applied to the restrictor by the injection port also enables split SFE-GC to analyze samples that, at least in our hands, could not be extracted either by off-line SFE or on-column SFE because of restrictor plugging. For example, the marine sediment described in Fig. 3 contains 2% elemental sulfur, which completely plugged the restrictor only a few seconds after off-line or on-column SFE was started. However, no plugging occurred with split SFE-GC, and the analysis could be performed directly on the sediment with no sample pre-treatment [8].

4.4 Summary

Despite being a newly-developing technique, the ability of SFE-GC to yield good chromatographic peak shapes and quantitative results has been demonstrated for a variety of analyte/matrix combinations. SFE-GC eliminates the need for sample handling between extraction and GC analysis, and reduces the total time required for extraction, concentration, and GC separation to less than one hour compared to the several hours or days required for conventional liquid solvent extraction/concentration techniques. On-column SFE-GC yields maximum sensitivity by transferring 100% of the extracted analytes to the capillary GC, but is best suited to dry samples smaller than 1 gram. Split SFE-GC can be used for larger samples, and a broader range of matrix types than on-column SFE-GC, including samples containing high concentrations of water and matrix components such as elemental sulfur. Future improvements in SFE-GC techniques, and better understanding of the mechanisms and kinetics that control SFE will doubtless increase the range of sample types and sizes that can be successfully analyzed using SFE-GC.

Acknowledgements. The author would like to thank the U.S. EPA, Office of Exploratory Research, and the New Jersey Department of Environmental Protection, Division of Science and Research, for support of his SFE investigations. Instrument loans from Suprex Corporation (Pittsburgh, PA, USA) and ISCO (Lincoln, NE, USA) are also appreciated.

References

[1] Nielen MWF, Sanderson JT, Frei RW, Brinkman UAT (1989) J Chromatogr 474:388
[2] Anderson MR, Swanson JT, Porten NL, Richter BE (1989) J Chromatogr Sci 27:371
[3] Liebman SA, Levy EJ, Lurcott S, O'Neil S, Guthrie J, Ryan T, Yocklovich S (1989) J Chromatogr Sci 27:118
[4] Wright BW, Frye SR, McMinn DG, Smith RD (1987) Anal Chem 59:640
[5] Levy JM, Guzowski JP, Huhak WE (1987) J High Resolut Chromatogr Commun 10:337
[6] Levy JM, Guzowski JP (1988) Fresenius Z Anal Chem 330:207
[7] Levy JM, Cavalier RA, Bosch TN, Rynaski AM, Huhak WE (1989) J Chromatogr Sci 27:341
[8] Hawthorne SB, Miller DJ, Langenfeld JJ (1990) J Chromatogr Sci 28:2
[9] Levy JM, Roselli A (1990) J High Resolut Chromatogr 13:418
[10] Hawthorne SB, Miller DJ (1987) J Chromatogr 403:63

[11] Hawthorne SB, Miller DJ, Krieger MS (1989) J Chromatogr Sci 27:347
[12] Hawthorne SB, Krieger MS, Miller DJ (1988) Anal Chem 60:472
[13] Giddings JC, Myers MN, McLaren L, Keller RA (1968) Science 162:67
[14] Hawthorne SB, Miller DJ, Krieger MS (1980) Fresenius Z Anal Chem 330:211
[15] Hawthorne SB, Krieger MS, Miller DJ (1989) Anal Chem 61:736
[16] Hawthorne SB (1990) Anal Chem 62:633 A

5 Gradients in SFC

ERNST KLESPER and FRANZ P. SCHMITZ

5.1 Overview

One of the features of supercritical fluid chromatography (SFC) which differentiates this type of chromatography from the neighboring chromatographic methods, gas chromatography (GC) and liquid chromatography (LC), is the larger number of gradients which may be useful in SFC. Gradients of temperature, pressure, density, linear velocity, and of the composition of the mobile phase, all greatly influence the chromatographic separation. So far, most often pressure and density have been programmed to form gradients, followed by temperature and composition gradients. Velocity gradients have been seldom used as single gradients, but they have been employed routinely in capillary SFC as a means of creating a pressure or density gradient. This combination of two gradients is, however, due to hardware limitations because the commercially available hardware for capillary SFC does not yet allow the programming of pressure without programming velocity at the same time.

Multiple gradients of the simultaneous type, where two or more independent programs are run at the same time, are now in the state of development. Particularly the simultaneous gradients of pressure and temperature, or of density and temperature, which appear to lead to higher resolution than single pressure or single density gradients, are being used more often. This applies to a lesser extent also to simultaneous gradients of composition and temperature. Of present day interest are furthermore the binary pressure-composition and the ternary pressure-composition-temperature gradients. The combined effect of the increasing pressure and of the increasing amount of the more powerfully dissolving component in the mobile phase may lead to a greater lowering of the capacity ratio of the analyte as compared to a pressure or composition gradient alone. This is specifically the case when a relatively non-polar primary component of the mobile phase, e.g. CO_2, is combined with a more polar secondary component, like acetonitrile, dioxane, or methanol, which has a greater solubilizing power than the primary component for a given analyte. The linear velocity may also claim a place in the realm of simultaneous gradients, because increasing density or increasing content of a more polar and higher molecular weight secondary component usually leads to lower interdiffusion coefficients of the analyte, other conditions being equal. The detrimental effects of lower diffusion coefficients may, at least in theory, be counteracted by lower linear velocities.

As opposed to simultaneous gradients, consecutive gradients have not yet found application in practice. It can be speculated, however, that the combination of consecutive temperature-pressure or pressure-composition gradients may become of practical value in the future. The first combination, for instance, may be employed to proceed from the realm of GC to that of SFC during a given chromatogram. Given suitable particle size in packed columns and suitable diameters in open capillary columns, the temperature gradient in the GC realm has the advantage of relatively high linear velocities at the van-Deemter-plot minimum, i.e. has the advantage of high analysis speed, even if the plate height may be larger than in the SFC domain. The pressure program, which follows the GC temperature program, leads to the SFC region. This region has the advantage of no longer relying solely on the vapor pressure of the analyte for transport in the column but also on the dissolution of the analyte by the compressed gaseous mobile phase. Therefore, the compounds in the analyte which possess no vapor pressure may also be moved through the column. The second combination of consecutive programs, pressure-composition, may be of value for analytes which are difficult to dissolve and to move through the column by an increase in pressure alone. Putting the pressure program first, and letting the composition program start later, has the advantage that the pressure program may already start at pressures too low for a secondary eluent to form a single gaseous phase with the primary eluent. Moreover, a pressure program, starting at a high free volume and with a low molecular interaction between the unpolar primary component of the mobile phase and the analyte, leads to higher interdiffusion coefficients. The consecutive composition program then introduces a more strongly solvating secondary component which dissolves even the components of the analyte which were not transported during the pressure program. In Table 1 the single and the simultaneous gradients are listed, and also examples of possible consecutive gradients.

For implementing suitable gradients it is of decided advantage to know at least approximately the interdependence between physical and chromatographic parameters for the system of interest, i.e. for a given type of triad: mobile phase, stationary phase and analyte. For the triad: mobile phase, CO_2; stationary phase, untreated silica; and an analyte consisting of a mixture of aromatic hydrocarbons (PAH) (naphthalene, anthracene, pyrene, and chrysene), enough data have been collected to present the interdependence between physical and chromatographic parameters by way of three-dimensional plots, showing the interdependence of two physical and two chromatographic parameters. The data for the three-dimensional plots have been obtained by series of isocratic-isobaric-isothermal chromatograms. While CO_2 in combination with untreated silica often shows a tendency for tailing and larger plate heights, it can be considered as one of the simplest combinations of mobile and stationary phase for SFC. The mixture of four aromatic hydrocarbons represents an analyte which is transported through the column both by vapor pressure and by solvation, the relative magnitudes of which depend on density and temperature of the mobile phase.

Table 1. Gradients in SFC

I Possible single gradients
1) Temperature
2) Pressure/density
3) Velocity
4) Eluent composition
II Possible multiple simultaneous gradients
A) Isocratic gradients
a) Pressure/density – temperature
b) Pressure/density – velocity
c) Temperature – velocity
d) Pressure/density – temperature – velocity
B) Non-isocratic gradients
Binary gradients
e) Eluent composition – pressure/density
f) Eluent composition – temperature
g) Eluent composition – velocity
Ternary and quaternary gradients
e.g. eluent composition – pressure/density – temperature
III Multiple consecutive gradients
Gradients of I and II following each other consecutively,
A) sequences of single gradients, e.g. temperature gradient followed by pressure gradient and pressure gradient followed by composition gradient
B) sequences of simultaneous gradients, e.g. pressure – temperature simultaneous gradient followed by composition – temperature simultaneous gradient

In Fig. 1, a plot of the capacity ratio of chrysene, k′(C), is shown versus the pressure at the column exit, p_e, and the column temperature, T [1, 2]. The average resolution between the four aromatic hydrocarbons, R_m^*, is shown by shading the three-dimensional surface of k′(C). There is a large peak for k′(C) at low p_e and T. Moreover, at a given p_e, the k′(C) run through maxima with rising temperature. In Fig. 2, the R_m^* is plotted instead of k′(C), while the shading remains to apply to R_m^* to show more clearly the height of the R_m^* surface [1]. Obviously, maxima are also observed for R_m^* in Fig. 2. However, the maxima for k′(C) and R_m^* do not necessarily coincide as is seen in Fig. 1 where R_m^* exhibits the highest values to the right of the peak on the three-dimensional k′(C)-surface, i.e. at higher temperatures. Several conclusions can be drawn from Fig. 1 for the choice of suitable pressure or temperature programs to be applied to this and similar chromatographic systems. For instance, R_m^* drops much more slowly with rising p_e and T than k′(C). Thus regions of p_e and T may be found where R_m^* is still sufficiently high for an analyte of many components but is combined with a k′ which is already tolerably low. Such regions of k′ are

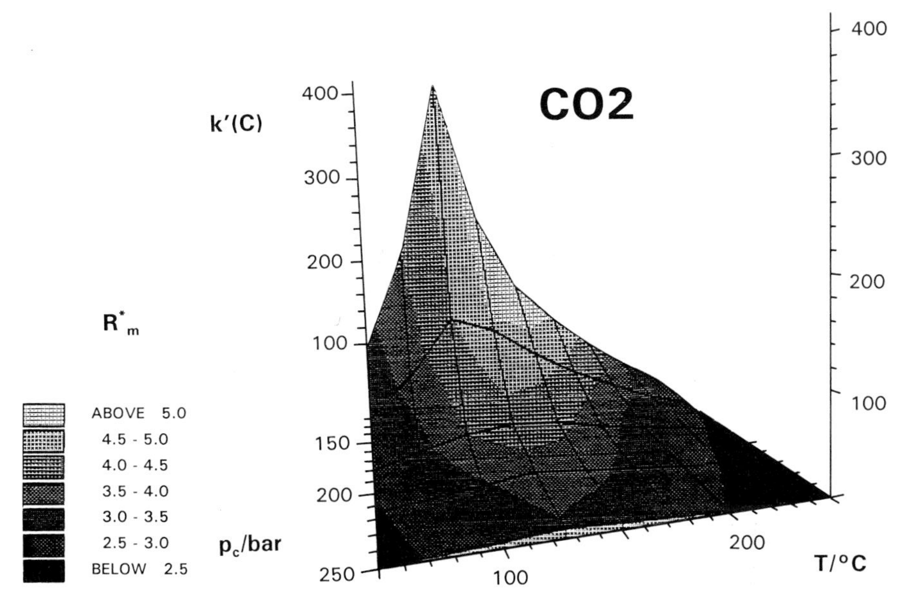

Fig. 1. Three-dimensional plot of the capacity ratio of chrysene, k′(C), versus column end pressure, p_e, and temperature, T. The average resolution between the components of the test mixture, naphthalene, anthracene, pyrene, and chrysene, R_m^*, is shown as color bands on the three-dimensional k′(C)-surface. The test mixture was separated on unbonded broken silica, Lichrosorb 100 (10 μm) (Merck, Darmstadt, FRG), in CO_2 as the mobile phase. The data have been collected by series of isobaric-isothermal runs, keeping the volume feed rate of the pump constant at 1 ml/min (liquid). Reprinted with permission [1]

preferably found at higher temperatures or higher pressures. If a pressure or temperature program is being run, one would preferably position as many program time increments, and corresponding k′(C) increments, in favorable regions as possible and also have the program start and end in one of these regions. If one intends to remain at the maximum of R_m^* throughout a positive pressure program, one would remain at the temperature of the crest of the R_m surface in Fig. 2.

Similar observations may be made for density programs. Inspecting the plot of Fig. 3 where the p_e-axis of Fig. 1 has been substituted by a density axis, i.e. a ϱ_e-axis, it becomes obvious that k′(C) decreases strongly both with increasing ϱ_e and increasing T. At the same time, the resolution R_m^* does not nearly decrease as much. It is seen, for instance, that there is a broad band for the range of $R_m^* = 3.0-3.7$ in the relatively low k′-region. Since the bands have the tendency to run diagonally across the k′(C)-surface, it may be of advantage to combine a negative T-program with a positive ϱ_e-program in order to remain at higher R_m^* throughout the program and the final density for the last eluting compound. If, however, the ratio $R_m^*/k'(C)$ is to be optimized throughout the program instead of R_m^*, a negative T-program combined with a positive ϱ_e-program might not be of advantage.

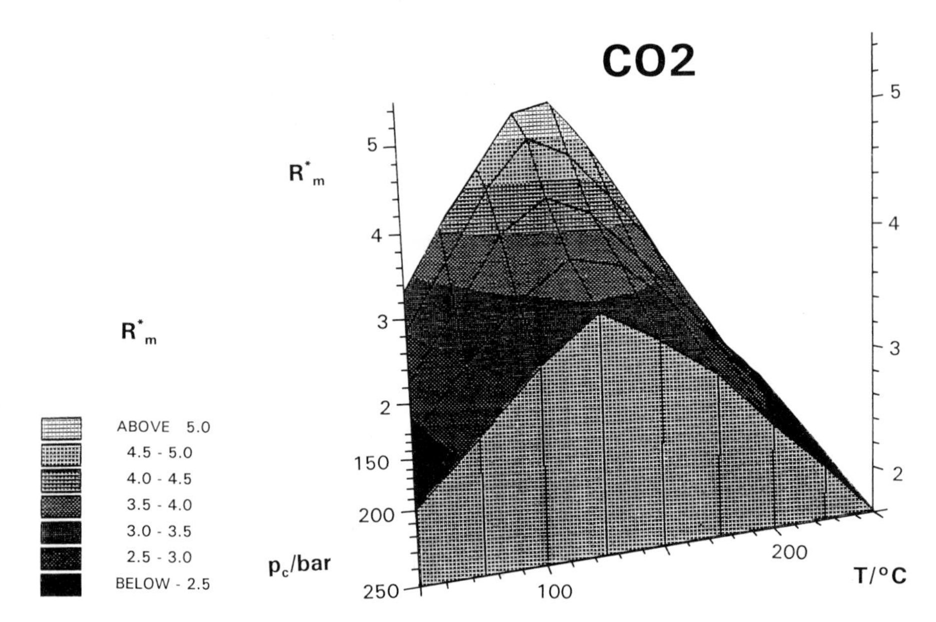

Fig. 2. Three-dimensional plot of the average resolution, R_m^*, versus p_e and T. For exhibiting more clearly R_m^*, the different ranges of average resolution are additionally presented as color bands. Other conditions as for Fig. 1. Reprinted with permission [1]

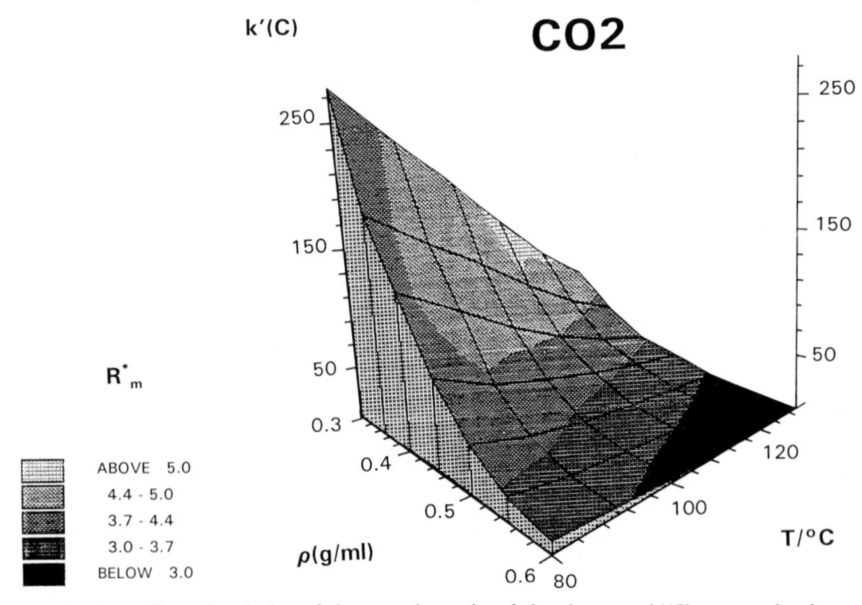

Fig. 3. Three-dimensional plot of the capacity ratio of the chrysene, k'(C), versus density at the column end, ϱ_e, and the temperature, T. The average resolution, R_m^*, is again seen as color bands. Other conditions as for Fig. 1. Reprinted with permission [1]

It remains to be seen to what extent the conclusions drawn from the data for the simple chromatographic system CO_2-silica-PAH apply to other chromatographic systems in SFC. It has to be considered thereby, that the volume feed rate of the pumps has been kept constant for the isobaric-isothermal chromatograms and, therefore, the linear velocity variable. So far, similar observations have been made, for instance, for pentane-silica-PAH [3–5] and for mixed mobile phases containing 1,4-dioxane, using dioxane treated silica and PAH [6–9]. For systems with high molecular mass analytes only a few, and for non-silica stationary phases no collection of systematic data has been reported so far.

For lack of space and, in part, also because of lack of published work it is only possible to treat more fully the single gradients of Table 1 (item I) but not the multiple gradients (item II) and consecutive gradients (item III). Even for single gradients, the eluent composition gradients have been omitted. The multiple simultaneous gradients [10, 11] and the multiple consecutive gradients [12, 13] have been treated in some research communication, and two brief reviews have appeared on the eluent composition gradients [14, 15]. In the following, the single gradients temperature, pressure, density, and velocity will be treated in turn. The main emphasis will be on literature which has appeared in the last years, because previous reviews on these gradients have already surveyed the earlier literature [10, 11, 14, 15].

5.2 Temperature Gradients

Referring to Fig. 1 again, it is obvious that k′ is reduced going to either higher or lower temperatures at a given pressure, provided one starts on different sides of the "crest" of the peak maxima for k′. The peak maxima at different pressures form a "crest" which runs from the very high crest at low pressures to the much lower crest at higher pressures (at 250 bar the crest is located at about 130 °C). The k′ is also strongly reduced when going to higher pressures at a given temperature. Similar observations are made with respect to R_m^* in Fig. 2 except that the crest for R_m^* remains much higher at higher pressures. There may be cases, however, where no maximum of k′ occurs with respect to temperature at a given constant pressure, but only a monotonous rise of k′ [16], or a more plateau-like behavior [17, 18]. In comparison to studies on k′, there exist still fewer systematic studies on the behavior of resolution, plate number, and selectivity for different chromatographic systems, i.e. different triads of mobile phase, stationary phase, and analyte. Therefore, no generally valid behaviors of the main chromatographic parameters can be set forth experimentally as yet. Provided a general behavior as in Figs. 1 and 2 is followed, both negative and positive temperature gradients lead to a reduction in k′ at a given p_e, if one starts the program at different sides of the crest which is formed by the k′ maxima. The concurrent reduction in resolution is not necessarily as severe percentagewise as the reduction in capacity ratio. Moreover, the reduction in resolution seems to be smaller with a positive versus a negative T-program, particularly at lower pressures. Inspecting Fig. 3, from which the behavior of k′ and R_m at a given density and temperature can be derived, it appears that only a positive temperature program at constant density leads to a reduction in k′(C).

Turning now to chromatograms obtained with actual T-gradients, instead of results obtained with series of isobaric-isothermal chromatograms, one may consider first negative T-programs. Negative T-programs provide a simple means of running a simultaneous, dependent, and positive ϱ_e-program, provided the pressure is not changed, or, at least, not decreased much during the negative T-program. In Fig. 4 the separation of a commercial wax with CO_2 on a capillary column is shown [19]. The temperature was programmed from 75 °C down to 33 °C, while the pressure was kept constant at 15.0 MPa. The increase in density during the run led to a chromatogram whose later eluting peaks became reasonably spaced, without the pro-

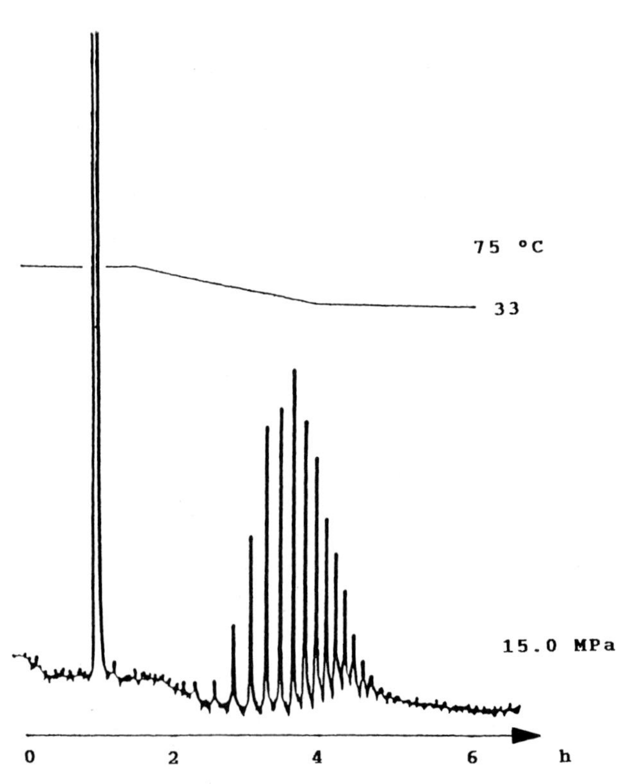

Fig. 4. Negative T-program at p = const. to yield a positive ϱ_e-program. Separation of a commercial wax in CO_2 on a capillary column (20 m × 100 μ) coated with BP1 bonded phase (0.1 μm film thickness). Reprinted with permission [19]

gressively wider spaced peaks usually observed in isobaric-isothermal runs of homologous series. In another instance, negative T-programming at constant column outlet pressure was employed to separate prepolymers of phenol-formaldehyde resins. The mobile phase was a mixture of CO_2 and ethanol and the stationary phase a silica-ODS bonded phase. Both novolac and resol oligomers could be separated up to at least eight phenol groups [20]. Negative T-programming at ϱ_e = const. instead of p_e = const. would probably lead to an increase instead of a decrease in k′ (cf. Fig. 3) for the eluting peaks.

Positive T-programming at constant pressure has not often been employed by itself in SFC, although the threedimensional graphs in Figs. 1 and 2 show for this and similar [21] chromatographic systems that positive T-programming may be of advantage in as much as capacity ratios are reduced to lower levels, while still maintaining more favorable resolution in comparison to the negative T-program. It has usually been preferred to rely on positive pressure or density programming instead of positive T-programming. It should be noted in this connection that pronounced maxima of k′ in dependence of T at p = const. are found both with bonded and unbonded silica packed columns and also with polymer coated open tubular capillaries [22].

Positive temperature differentials between sample injector valve and column, e.g. keeping the sampling valve at 25 °C and the column at 80 °C at low pressures of 10 MPa for CO_2 as the mobile phase, has led to sample trapping at the column inlet. In this way broadening effects caused by the sampling procedure, including infavorable flow characteristics and too large an amount of sample solvent, are reduced greatly [23]. Afterwards a temperature program for the column might start at 80 °C. Of vital interest for temperature programming is also that bonded phase silica may change its chromatographic properties at elevated temperatures. While, for instance, silica-ODS (C-18) columns are stable up to 150 °C and may even improve upon conditioning at higher temperatures, the APS (aminopropyl) columns are already unstable at temperatures as low as 75 °C [24].

5.3 Pressure Gradients

The maxima in resolution found in dependence of T at p = const. have repeatedly been mentioned in the previous sections. A related phenomenon can be observed if an identical pressure gradient is run at a series of different temperatures. In Fig. 5 the same pressure program is repeated at six different temperatures from 50° to 170 °C for a polyethylene glycol analyte in CO_2 on an open tubular capillary coated by a polysiloxane. As judged by visual inspection, the resolution increases up to temperatures of 100° or 125 °C and appears to decrease again at still higher temperatures [25].

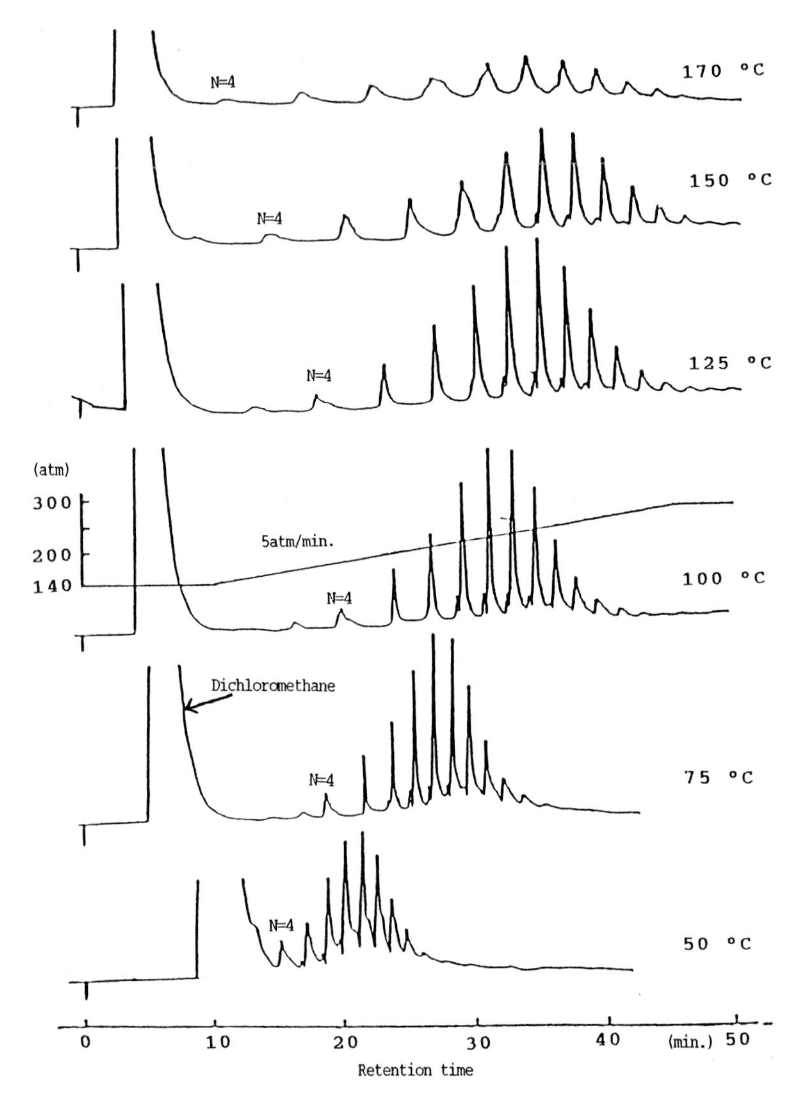

Fig. 5. Runs of the same pressure program at different temperatures, starting with 10 min hold time at 140 bar and then rising to 300 bar at a rate of 5 atm/min. FID temperature 350 °C. Analyte is poly(ethylene glycol) dissolved in CH_2Cl_2 (4.4%). Capillary column (7 m × 100 μm i.d.) wall coated by poly(cyanopropyl dimethyl-siloxane). Reprinted with permission [25]

Apparently, there exists a temperature range which leads to a maximum in resolution not only at p = const. but also when employing a pressure gradient. This conclusion can also be derived from Fig. 2. Besides using the optimum temperature range for pressure programming, one may also consider to adjust the nature and the composition of the mobile phase. Thus instead of pure CO_2, a binary mobile phase consisting of CO_2 and 2-propanol has been used. In this way, the desired lower capacity ratios and lower retention times for a given pressure gradient have been obtained by way of higher polarity [26, 27]. Moreover the possibility exists that a specific composition of a binary mobile phase leads at a given temperature and pressure range to a maximum in resolution which may exceed the resolution obtainable with the pure primary component [28].

Examples of employing pressure gradients, past and recent, are very numerous. Pressure gradients have been used frequently, for instance, for oligomer separation, e.g. for oligo (methylphenylsiloxane) [29], and relatedly, also for polymer additives [30]. The ramp of the pressure gradient should be generally chosen to possess a steepness which is suited for obtaining a good, or even optimal resolution for a given packed column [31, 32] and the same should apply to OTC columns. As has been experimentally found, the pressure ramps may be made very steep and one may still retain resolution which is sufficient for a specific analyte. This has been demonstrated for short OTC columns of small diameters of 25 and 50 μm. Figure 6 shows a chromatogram obtained with a ramp of 35 atm/min completed within a few minutes [33]. The ramp could be increased to 100 atm/min, still retaining reasonable resolution. Thus increased ramp speed and decreased OTC column length may be used to speed up separations greatly. One may utilize this finding to shorten the time for method development [34]. After a suitable method has been found, a longer capillary and a slower ramp can

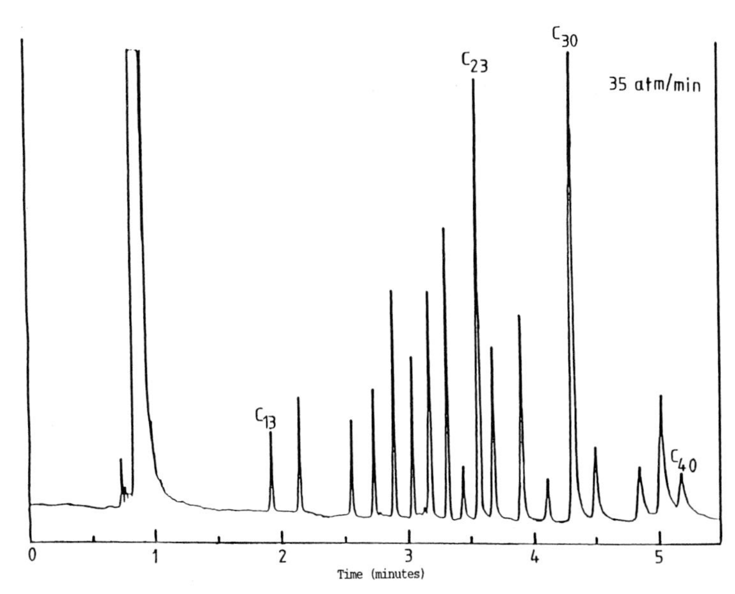

Fig. 6. Rapid pressure ramp of 35 atm/min at an initial pressure of 75 atm for CO_2 at 100 °C, as employed for the fast separation of an alkane mixture on a 1.5 m×25 μm i.d. capillary fused silica column coated with a 0.15 μm film of SE-54 (5% phenyl in a poly(methylphenylsiloxane). Reprinted with permission [33]

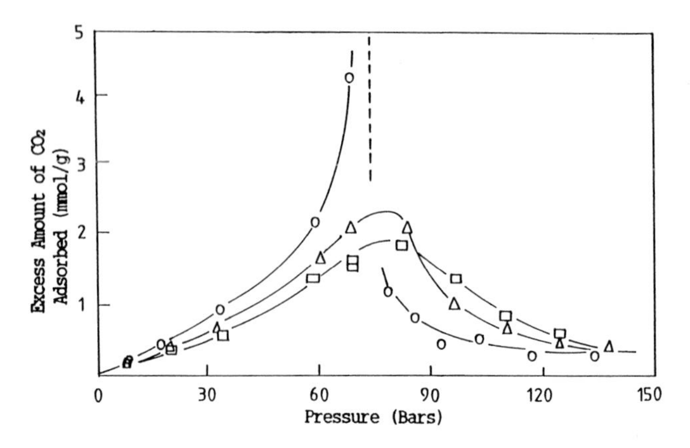

Fig. 7. Surface excess (Gibbs) adsorption isotherms of CO_2 on octadecyl-bonded silica at three different temperatures: (○) = 30 °C; (△) = 40 °C; (□) = 50 °C. The *dotted line* is the vapor pressure of subcritical CO_2 at 30 °C. Reprinted with permission [35]

be applied to increase resolution as much as is needed for difficult analytes.

With both solid adsorbents and liquid absorbents, it is of interest to know how the sorption of the mobile phase on, or in, the stationary phase changes with temperature, pressure, or density. It has been found in this respect that the Gibbs surface excess adsorption on solid adsorbents runs through a maximum with pressure. With rising temperature, on the other hand, there appears to be a decrease of adsorption at lower pressures and an increase at higher pressures [35]. In Fig. 7 the surface excess adsorption isotherms of CO_2 on silica-ODS are shown at three different temperatures. The maxima of the isotherms are clearly seen and also the changed relative positions of the isotherms for the different temperatures at pressures below and above the maxima. The changed relative positions represent the temperature influence [35]. Results pertaining to absorption in a polymeric stationary phase have been obtained earlier and apparently differ from the results obtained for adsorption [36, 37]. Another study was concerned with the uptake of CO_2 and *n*-butane by polymeric (liquid-like) stationary phases. A large uptake was found for both fluids at supercritical conditions which was independent of density above 0.2 g/ml [38]. In this connection, it is of practical interest how fast such adsorption and absorption processes proceed and what the minimum elution volume requirements are to reach equilibrium again after the pressure/density or the composition of the mobile phase is changed. In SFC, a pressure change was found to require less than two column elution volumes and a change in modifier 10 to 30 column elution volumes. In HPLC, a change in modifier for HPLC takes one order of magnitude more, or even longer [39]. The concept of the threshold pressure may be of value to determine the minimum pressure below which a pressure program should not start. The threshold pressure, which may be looked upon as the pressure at which the solubility of a given compound starts to be significant in a supercritical fluid, may be determined by simple theory based on the Hildebrand solubility parameter and the Flory Huggins interaction parameter, both as adjusted to supercritical fluids [40].

Much attention has been paid recently to a hardware device which is essential to perform a pressure gradient. This is the backpressure device downstream of the column, which allows to program the pressure during a separation. Programmable backpressure devices for larger diameter packed columns as well as for capillary columns have already been proposed earlier. However, the demands on precision put forward by the low flow rates in small diameter columns, particularly open tubular capillaries of 50–100 μm diameter, are stringent. For the higher flow rates of larger diameter packed columns automated valves have recently been employed routinely, whereby the valves are preferably part of a feedback loop. These newly developed regulating valves are based on high speed flow switching, that is, on a valve action of continuously and quickly opening and closing, thereby regulating the pressure upstream of the valve [41, 42]. The continuous flow switching has the purpose to prevent the clogging of the valves by accidentally occurring particulate matter, which clogging occurs more easily when the gap between needle and seat of the valve is smaller by keeping the valve constantly open. Automated needle valves have not found much use for capillary columns so far because of the very low flow rates of such columns which place still more stringent requirements on the precision of the valves. Instead, the effluent of the column (or of the detector following the column) is led into an auxiliary stream of liquid, whereby it is the pressure of this auxiliary stream which is monitored and programmed. The pumping device which creates this auxiliary stream may either be volume flow controlled, or more elegantly, pressure controlled. In any case, a resistor is needed at the outlet of the combined streams, which is usually a simple restrictor, that is a piece of capillary or modified capillary. A simplified scheme of a SFC apparatus illustrating the principle is given in Fig. 8 [43]. Instead of a make-up liquid as the auxiliary stream a make-up gas has also been proposed. This has been put into practice by a so-called sheath-flow nozzle shown in Fig. 9 [44]. The make-up gas, argon, enters from the side and exits through a laser drilled orifice together with the effluent from the capillary column. When the make-up gas is pressure controlled, this controls then also the pressure at the column exit.

The back pressure devices discussed allow to control the volume or mass flow rate of the pump independently from the pressure. The linear velocity in the column, and to some extent the pressure drop over the column, is then also independent from the pressure at the column outlet. This independence is not available with the commonly used restrictors for capillary columns which are invariable back pressure devices, represented by, for instance, a simple piece of small diameter capillary tubing, or a pin hole, or a frit at the end of a capillary tubing. There have been efforts to change

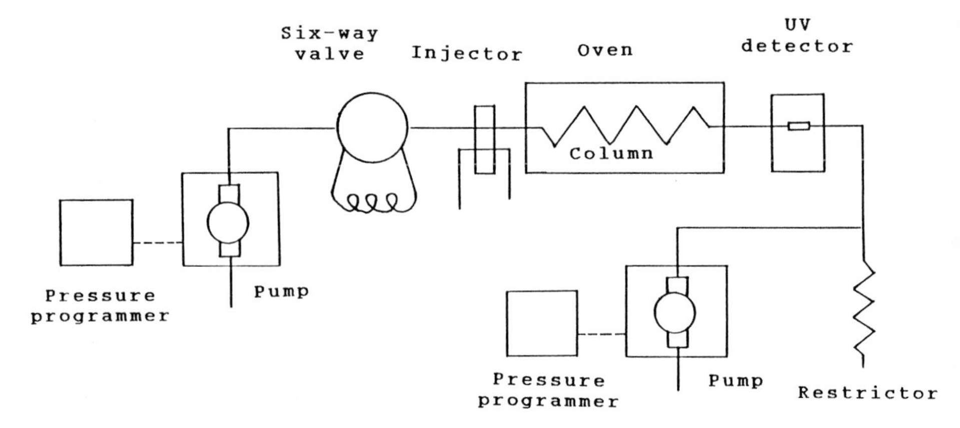

Fig. 8. Schematic diagram of an SFC apparatus with an auxiliary stream of liquid at the column (detector) outlet. The second pump delivering the auxiliary stream may be pressure programmed. The combined streams from the second pump and the separation column exit pass through a restrictor (a capillary of 200 μm i.d., 5–20 cm length, packed with 5 μm particles). Reprinted with permission [43]

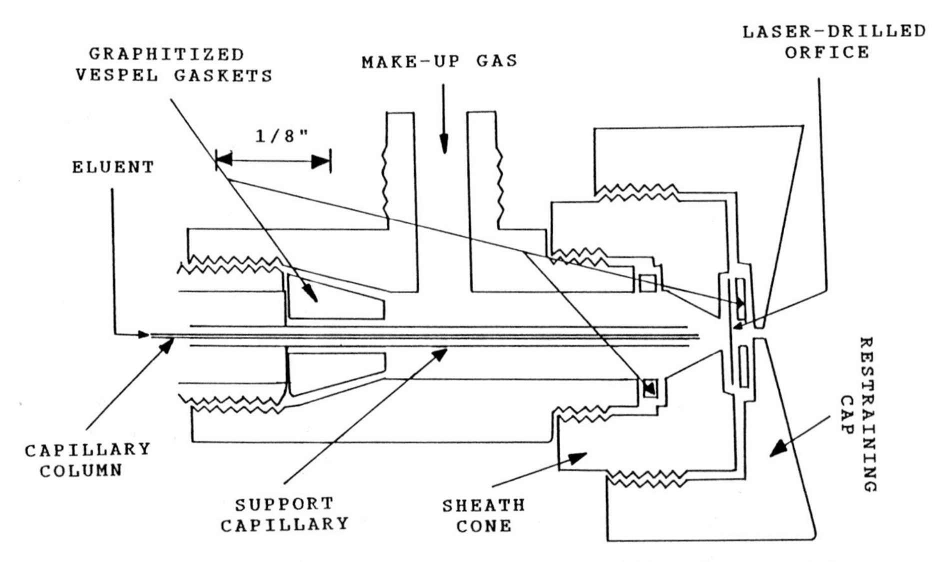

Fig. 9. Schematic drawing of the sheath-flow nozzle used as a variable back pressure device. For explanation see drawing and text. Reprinted with permission [44]

such invariable to variable resistors. A recent effort relied on changing the temperature of the capillary restrictor, because a higher temperature of the restrictor leads to a lower mass flow rate and lower linear velocity in the column at a given pressure [45]. In this connection it is of importance that the flow characteristics through capillary restrictors has been the subject of several basic studies [46–49].

5.4 Density Gradients

Density gradients are implemented by programming the pressure in such a way as to program the density linearly, asymptotically, or in any other desired way. Toward this end, the relationship between pressure at a given temperature on one hand, and the density on the other, must be known. Since this relationship is not experimentally determined in full for all single (pure) mobile phases and even less so for the binary or ternary mobile phases of potential interest for SFC, approximative calculations may have to be carried out [50]. The reason for employing density controlled gradients instead of pressure controlled gradients is connected to the approximately valid equation:

$$\ln \langle k' \rangle_t = a - b \langle \varrho \rangle_t \quad (1)$$

where a and b are constants, and $\langle k' \rangle_t$ and $\langle \varrho \rangle_t$ are the temporal average capacity factor and the temporal average density over the column residence time, respectively. Thus there exists a simple dependence of retention on density, while the dependence of retention on pressure is considerably more complicated. To obtain a simple, although specific type of density program for the mobile phase, whereby the density increases progressively slower with time, one may simply program the pressure linearly with time above $T_r \approx 1.05$ (T_r = reduced temperature) whereby this density program tends to become more linear the higher the temperature range and the starting pressure of the program is chosen. The actual shape of these density programs may be obtained by referring to the common p_r – versus ϱ_r – isotherms (r = reduced). Equation 1 indicates that $\ln \langle k' \rangle_t$, which is the actually measured ln k′, may be plotted as a nearly straight line versus $\langle \varrho \rangle_t$ or ϱ, as has often been confirmed experimentally. Recently is been pointed out that the reciprocal retention time, t_R^{-1}, of some aromatic analytes, when plotted versus density of a supercritical pentane-3%-isopropanol mobile phase, might also yield straight lines [51].

A positive pressure program linear with time, leading to a density program with progressively slower increases in density, has been used in SFC-FID and SFC-MS mode to analyze oligomeric materials of low volatility or of thermally labile materials on OTC columns. As an example, Fig. 10 shows the SFC-FID of two series of ethoxylated alcohols mixed in the same chromatogram [52]. A similar analysis has been performed by an asymptotic density program on ethoxylation products of a mixture of 2-ethylhexanol and n-octanol. The SFC-FID response factors of the individual species of the ethoxylation product have been determined and the amounts of the individual species evaluated [53]. Higher molecular weight hydrocarbons have also been recently separated by density programming on OTC columns. Thus a separation of 2-vinylnaphthalene oligomers by a program consisting of three linear sections has been reported [54]. The combination

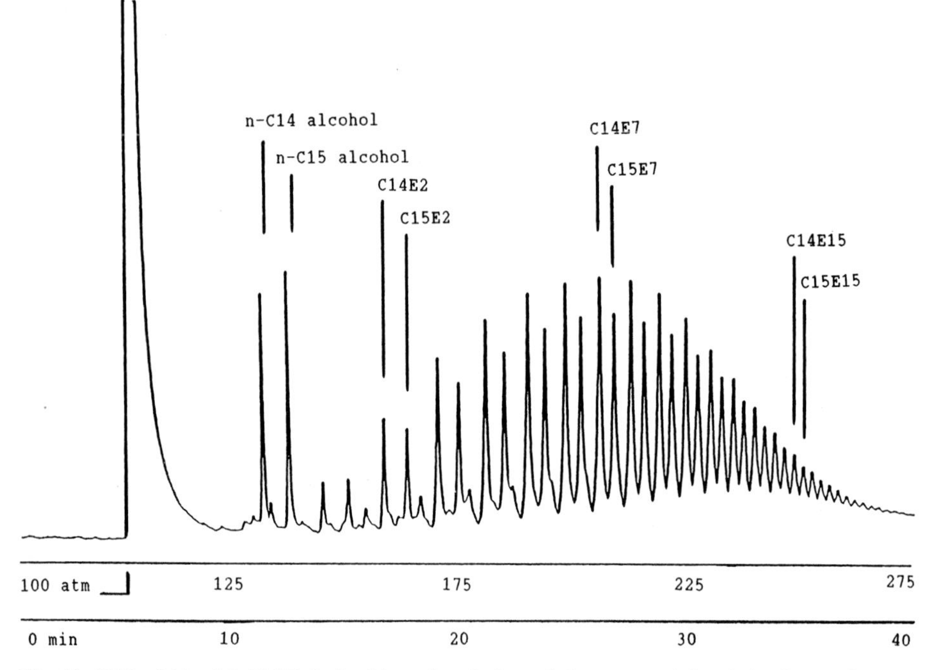

Fig. 10. SFC of Neodol 45-7T derived by ethoxylation of the two straight chain C_{14}- and C_{15}-alcohols. The peaks are identified by symbols where C14 E7 means, for instance, a molecule derived from the *n*-C_{14}-alcohol having a chain of 7 ethyleneoxide units. The Neodol 45-7T sample has been "spiked" by adding *n*-C_{14}- and *n*-C_{15}-alcohols for identification purposes. Reprinted with permission [52]

of supercritical fluid extraction (SFE), open tubular capillary SFC, and detection by FT-IR microspectrometry has been utilized for the separation and identification of polycyclic aromatic hydrocarbons in a coal tar pitch [55]. The linearly density programmed capillary chromatogram is seen in Fig. 11. Despite the complicated multi-species mixture, possessing a wide range of retentions, the components of the pitch are relatively well resolved. SFE extraction steps, using CO_2 at different densities as the extraction medium, may be used to provide an analyte for SFC containing a reduced number of components, as is indicated on top of Fig. 11.

Besides the application of SFC to higher molecular weight analytes, thermolabile substances are also often analyzed by SFC. An example is the separation of isocyanates and derived compounds like dimer, trimer, urea, biuret, carbodiimide, carbamate, thiocarbamate and others. By employing density gradients, it has been found that SFC is a very suitable, if not superior technique for these classes of compounds [56]. In Fig. 12 the density programmed separations of a number of diisocyanates is shown isothermally at 50, 75 and 100 °C in CO_2 on an OT-capillary column. The peak widths decrease with increasing temperature and the same applies to the

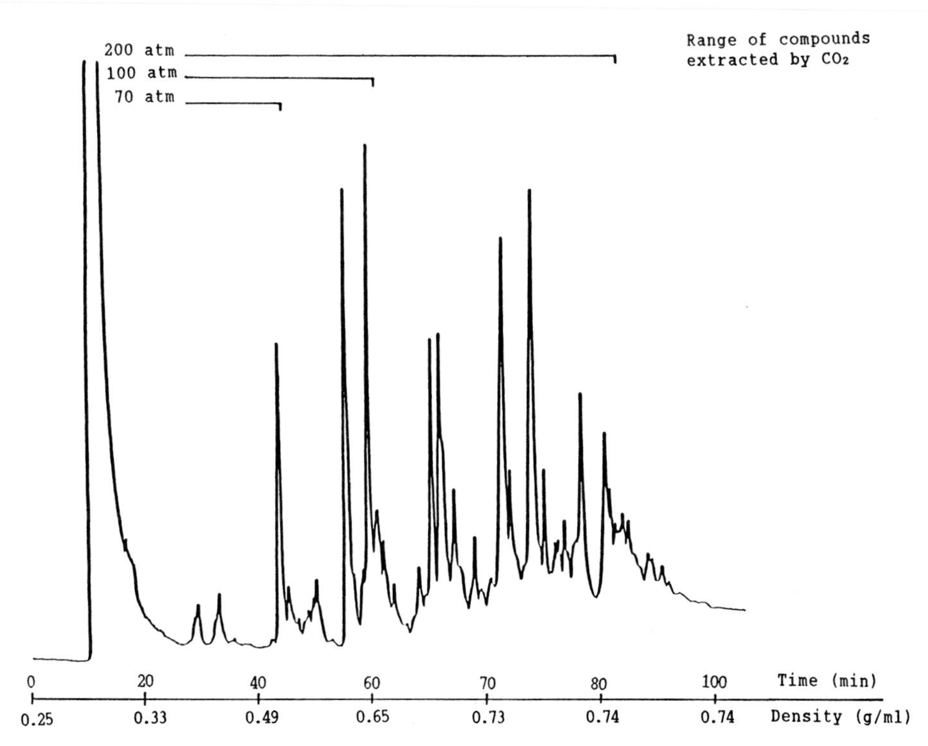

Fig. 11. SFC of a dichloromethane soluble fraction of coal tar pitch in CO_2 as mobile phase on a 10 m × 50 μm SB-Biphenyl-30 column (d_f = 0.25 μm) at 110 °C. Linear density programmed from 0.25 g/ml to 0.74 g/ml at 0.006 g/ml/min, ending with an isopycnic period of 20 min. The range of compounds extracted in an SFE step by CO_2, instead of extracting with dichloromethane, is given on top. Reprinted with permission [55]

retention times with the particular conditions of density programming used. A number of pesticides of the carbamate, chlorinated, and organophosphate type have also been chromatographed by density programming at different temperatures in CO_2 on capillary columns [57]. The same was reported for volatile nitrosamines [58]. Density programming has also been extended to multidimensional chromatography of the type: capillary SFC-capillary SFC [59] and liquid size exclusion chromatography-capillary SFC [60].

Several studies have been reported for determining the adsorption and absorption of single and binary mobile phases in dependence of density in SFC. In a recent communication, a brief summary of previous work is given and the excess Gibbs adsorption isotherms reported for the pure mobile phase CO_2 on unbonded and bonded silica gel [61]. The isotherms show very similar shapes for the different silicagels. The excess absorption, i.e. the absorption beyond the density of the bulk mobile phase, is found to possess a maximum near the critical pressure, p_c. The maxima are particularly pronounced near the critical temperature, T_c, and progressively flatten at

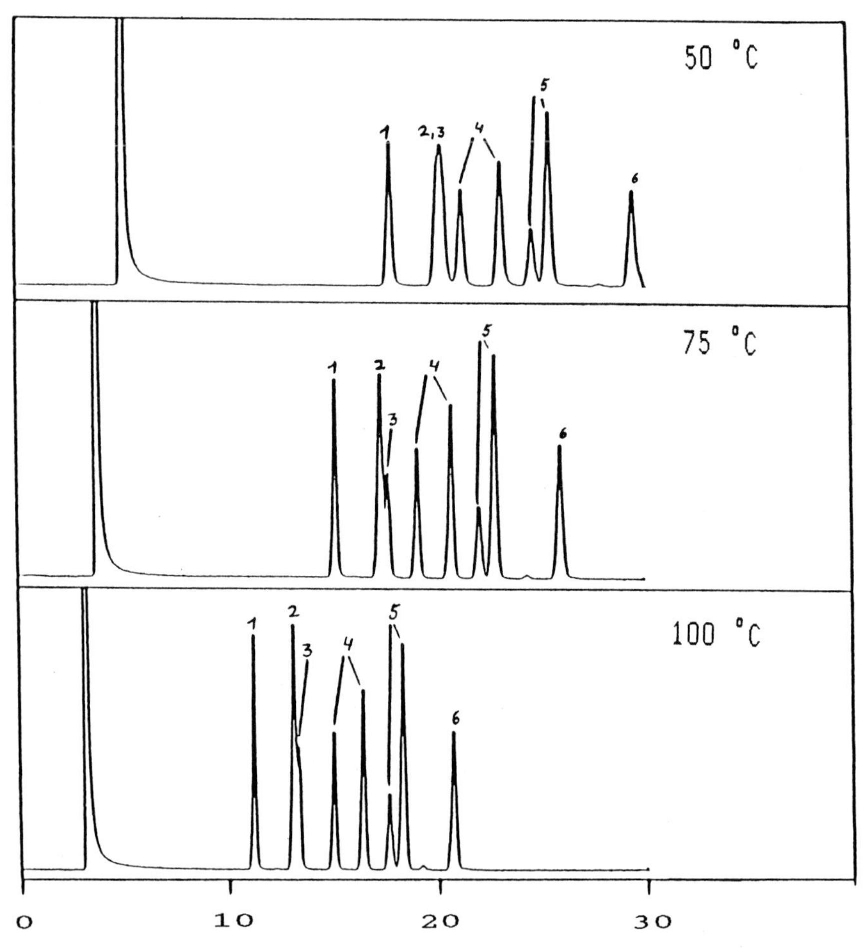

Fig. 12. Linear density programmed separation of a series of low molecular weight diisocyanates at three different temperatures on an OTC column (5 m×100 μm i.d. fused silica) coated with a poly(siloxane). *Peak 1:* *p*-phenylene diisocyanate; *peak 2:* 2,4- and 2,6-toluenediisocyanate; *peak 3:* 1,6-hexanediisocyanate; *peak 4:* 2,2,4-trimethylhexane-1,6-diisocyanate; *peak 5:* isophorone diisocyanate; *peak 6:* *m*-xylylene diisocyanate. Density program: 5 min initial density hold time (initial densities of 0.25 (50 °C), 0.18 (75 °C), and 0.15 g ml^{-1} (100 °C)), 0.01 g ml^{-1} min^{-1} ramp rate up to a final density of 0.5 g ml^{-1}. Reprinted with permission [56]

higher temperatures. In Fig. 13 the isotherms at 40 °C for CO_2 on several bonded SFC adsorbents are shown. Because for a monolayer the excess adsorption is about 9 μmol/m^2, there must exist a multilayer adsorption near the critical density ϱ_c (CO_2) = 0.466. The isotherms apparently are not much different for the three bonded phases and at low, subcritical densities the isotherms are even approximately linear. As an explanation for the appearance of a maximum one may offer that at moderate densities, i.e. below

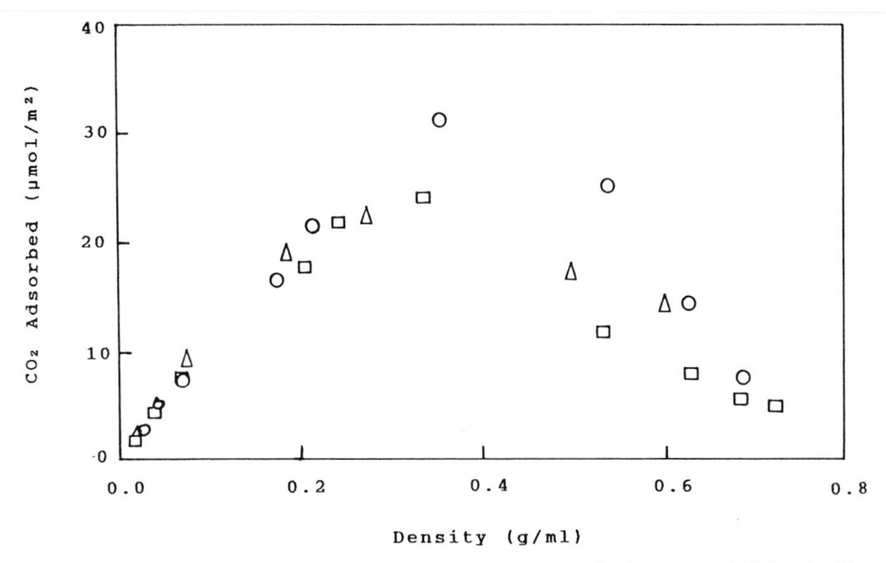

Fig. 13. Excess adsorption isotherms of CO_2 on several bonded silica gels at 40 °C. □ = C_{18} bonded silica, ○ = cyano bonded silica, △ = diol bonded silica (for further explanation see text). Reprinted with permission [61]

the adsorption maximum, the adsorption increases with density because of the increasingly higher bulk concentration of mobile phase molecules. Above the adsorption maximum, the bulk concentration of the mobile phase becomes already high enough for this phase to become a progressively better "solvent" for the adsorbed layers, redissolving and removing more and more of the excess adsorbate.

There is by necessity a pressure drop of the mobile phase over the length of a packed or of an open tubular column, even if this pressure drop is only small. There is also a corresponding density drop and a velocity increase, as seen from the inlet to the outlet of the column. The pressure and density drops are easily interconverted if the pVT-data are known, while the density drop yields the velocity increase directly. The changes of these three physical parameters over the length of the column are non-linear, in principle, although they become more linear as the compressibility of the mobile phase decreases. Decreasing compressibility is found in the region of higher densities and temperatures. A generalized treatment of the density over the column has been published [62]. Starting from Darcy's law, equations are derived for the average density, average retention time, and average capacity factor, whereby the spatial and temporal averages are distinguished. For instance, the temporal-averaged capacity factor $\langle k' \rangle_t$, which is the experimentally observed capacity factor, is given by

$$\langle k' \rangle_t = \frac{t_R - t_0}{t_0} \int_{\varrho_e}^{\varrho_i} k' D_t(\varrho) d\varrho \left[\int_{\varrho_e}^{\varrho_i} Dt(\varrho) d\varrho \right]^{-1} \quad (2)$$

where

$$D_t(\varrho) = \eta^{-1}\varrho^2(\partial p/\partial \varrho)_T \quad (3)$$

with t_R the retention time, t_0 the dead time, ϱ the density, ϱ_i and ϱ_e the densities at the inlet (i) and the outlet (e) of the column, η the viscosity, $(\partial p/\partial \varrho)_T$ the density coefficient of the pressure, and T the temperature. A calculation of η^{-1} $(\partial p/\partial \varrho)_T$ in dependence of ϱ at a given T via the Jacobsen-Stewart modification of the Benedict-Webb-Rubin equation of state [63] allows also to calculate the density profile over the column. With CO_2 at 320 K, a packed column inlet pressure of 120 bar, and an outlet pressure of 100 bar, the arithmetic mean density is reached not at 50% but at 63.6% of the column length, as counted from the column inlet [62]. An equation analogous to Eq. 2 and also an analogous equation for the spatial average of the density has been derived by other authors for a pressure instead of a density integral [64]. In Fig. 14 the spatial average of the density, $\langle\varrho\rangle_x$, is plotted as the percentage deviation from the density at the mean pressure $\varrho(p_{mean})$ ($p_{mean} = (p_i + p_e)/2$; p_i and p_e are the pressures at the inlet (i) and the outlet (e)). If the pressure drop is large, the deviations are also large, provided one is near the critical temperature and within the pressure range seen in Fig. 14 [64]. This finding is to be taken into consideration for theoretical work in as much as the spatial average of the density for a column which exhibits a larger pressure drop may not even nearly be set equal to the density at the mean pressure.

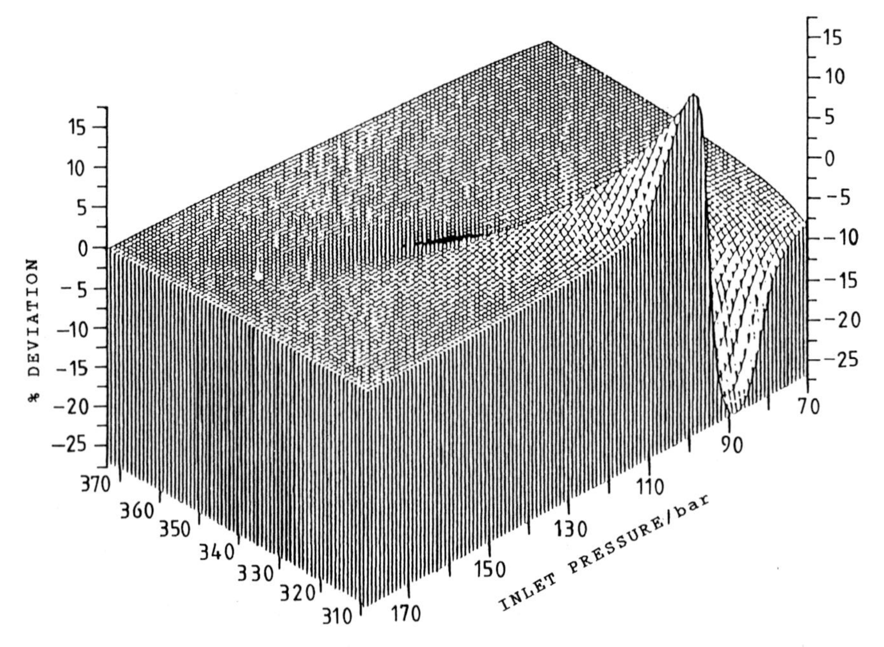

Fig. 14. Percentage deviation of the calculated spatial average density $\langle\varrho\rangle_x$ from the density at the mean pressure, i.e. 100 $[\varrho(p_{mean}) - \langle\varrho\rangle_x]/\langle\varrho_x\rangle$ at a pressure drop of 30 bar as a function of inlet pressure and column temperature. Mobile phase CO_2. Reprinted with permission [64]

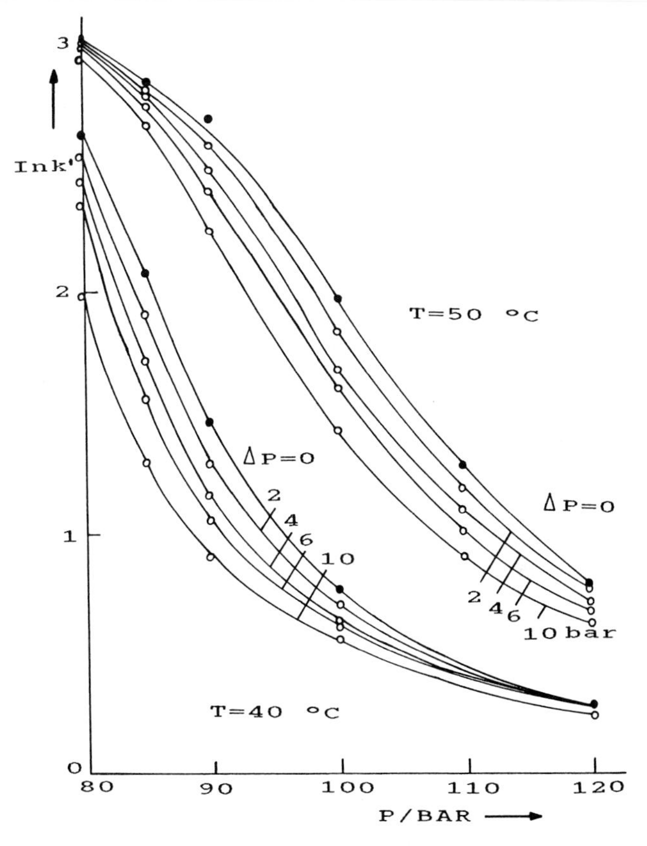

Fig. 15. Logarithm of the capacity ratio, ln k', versus mean column pressure, p_{mean} (= P in Fig. 15). Different curves correspond to different column pressure drops, Δp. The curve of $\Delta p = 0$ has been extrapolated from the experimental data obtained at different pressure drops. Reprinted with permission [65]

To what extent the experimentally found capacity ratio, $\langle k' \rangle_t$, on a column with pressure drop may vary from the capacity ratio on a column without pressure drop may be seen in Fig. 15. There the arithmetic mean pressure, p_{mean}, is plotted versus $\ln \langle k' \rangle_t = \ln k'$ for two temperatures of 40° and 50 °C [65]. The pressure drop $\Delta p = p_i - p_e$ is plotted as parameter. The different pressure drops have been obtained by varying the column outlet pressure at different constant inlet pressures and therefore by varying the linear velocity in the column. The column will yield the limiting capacity ratio at a given pressure when $\Delta p = 0$. At all higher pressure drops the capacity ratio at the same p_{mean} will be increasingly smaller. Because for open tubular capillary columns the pressure drop is usually small (<3 bar), the difference between the experimental capacity ratio of these columns and the capacity ratio one would obtain with a pressure drop of zero are only small and may be neglected [66].

5.5 Velocity Gradients

Gradients of velocity have been rarely used by themselves. Only a few examples of continuous [67] and stepwise [68] pressure programs have been reported. Thus the usefulness of stand-alone velocity gradients has not been demonstrated sufficiently in practice. However, van Deemter plots of plate height versus linear velocity have been obtained repeatedly for SFC. The behavior of the plots was recently studied again in dependence of pressure and the nature of the analyte, and also in comparison to GC and HPLC. Some conclusions as to the usefulness of velocity programming may be derived from these plots. In Fig. 16 the van Deemter plot in the form of reduced plate height versus reduced velocity is shown at three different

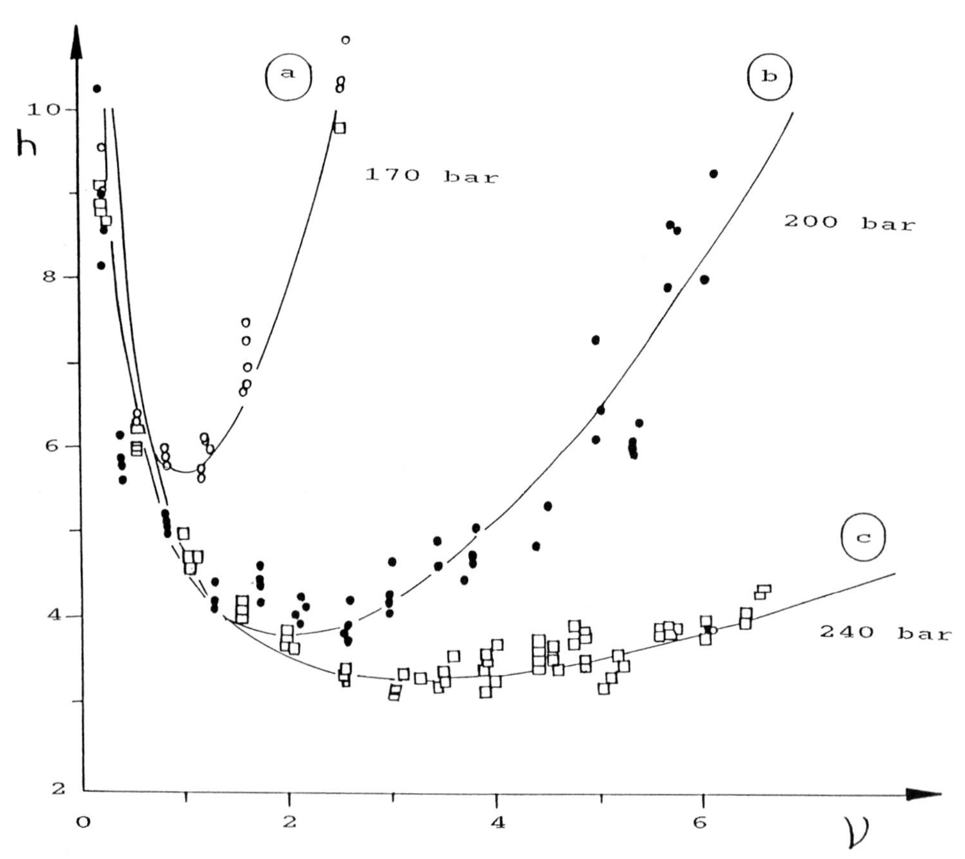

Fig. 16. Reduced plate height, h, versus reduced linear velocity of the mobile phase, ν, for phenanthrene in CO_2 at T = 50 °C at three mean column pressures (a) = 170, (b) = 200, (c) = 240 bar. Column 15×0.46 cm i.d., packed with 5 μm octadecyl bonded silica. Reprinted with permission [69]

pressures for phenanthrene in CO_2 on a column packed with bonded silica. Remarkably, the higher the pressure, the lower is the position and the slope of the high velocity branch, although the opposite may be expected on account of the interdiffusion coefficient decreasing in the mobile phase with increasing pressure. This phenomenon has been said to be connected to the difference in the capacity ratio at the inlet and the outlet of the column [69]. One may also consider other explanations: for instance, that the decrease is connected to the general decrease of k′ at higher pressures or to the type of porosity of the stationary phase. While the pore size of the 5 μm particle diameter octadecyl bonded silica has not been specified [69], the pore size will probably be small. Better accessibility of the pore walls for the analyte, because of a less extensive and more easily replacable coverage by CO_2 at the higher pressure of the mobile phase, may lead to more rapid adsorption and desorption at the surface of the pores. By the same token, interdiffusion more deeply into the pores will be hindered at higher densities of CO_2 in the pores with the result that mass transfer resistance is reduced. With respect to gradients, this data of Fig. 16, whenever valid in SFC, would mean that with positive pressure/density programming the efficiency becomes less sensitive to velocity increases at higher pressures. This would be of advantage because

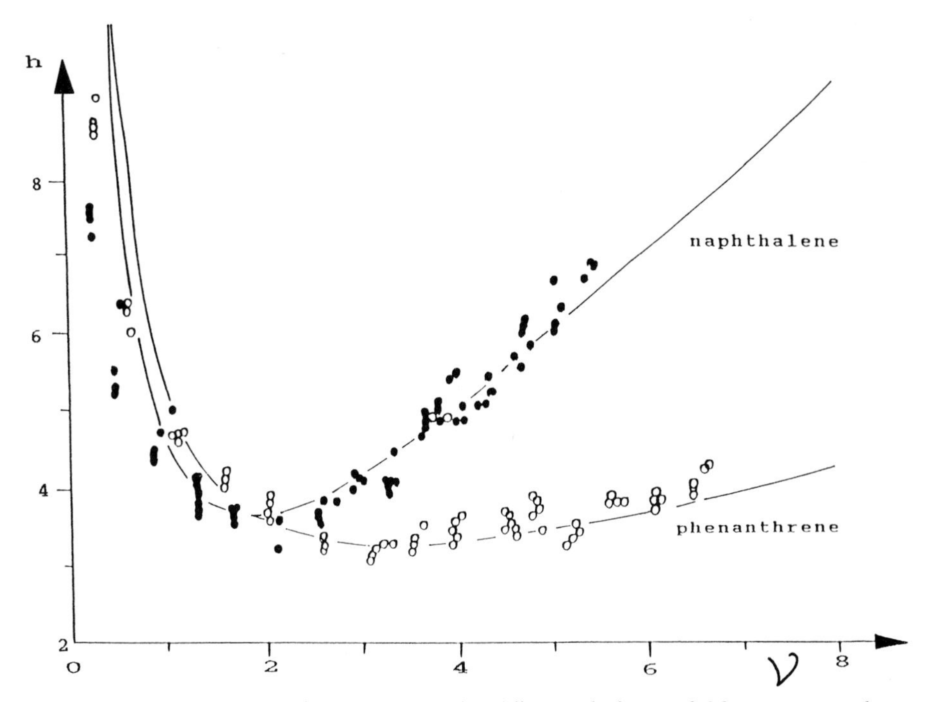

Fig. 17. Reduced plate height, h, versus reduced linear velocity, ν, of CO_2 at a mean column pressure of 240 bar and a temperature of 50 °C. Packed column as for Fig. 16. Reprinted with permission [69]

the reduction of the interdiffusion coefficient with increasing density during the program might be counteracted by this effect.

The molecular weight of the analyte proves to be of great influence also. Using the highest pressure of Fig. 16, i.e. 240 bar, the reduced plate height versus reduced velocity is plotted for phenanthrene again in Fig. 17, but is now compared to that of naphthalene under the same conditions [69]. The lower molecular weight compound naphthalene exhibits the steeper high velocity branch, despite the reasonable expectation that naphthalene possesses a larger interdiffusion coefficient than phenanthrene. This order in plate height for analytes of different molecular weights on a similar stationary phase has been observed before [70, 71]. It may be suspected from Fig. 17 that a positive velocity gradient at constant pressure, starting with a low velocity and increasing the velocity while the higher molecular weight components of the analyte are being eluted, might be a suitable procedure to reduce the analysis time, without much reduction in efficiency. It should be mentioned, however, that earlier results in SFC, obtained with stationary phases typical for packed column GC, showed a reversed order in height and slope of the high velocity branch both with respect to pressure and to molecular weight dependence [17, 72].

The comparison of SFC with GC and HPLC is shown in Fig. 18 as a plot of plate height H versus linear velocity u, employing columns packed with bonded silica [73]. The analytes were in all three cases compounds which were little or not retained, i.e. of a k′ approximating zero. The intermediate position of SFC between GC and HPLC is seen immediately in this van

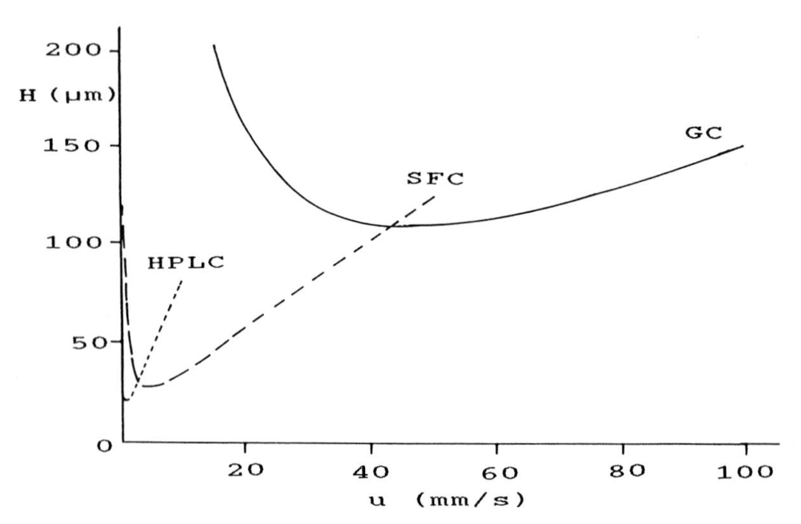

Fig. 18. Plate height, H, versus linear velocity, u, in GC, SFC, and HPLC. Column, 25×0.4 cm i.d., packed with RP-18, 10 μm. GC: mobile phase N_2; analyte, methane. SFC: mobile phase CO_2; analyte, methane. HPLC: mobile phase methanol; solute, benzene. Reprinted with permission [72]

Deemter plot, whereby the SFC-curve may be moved more towards the GC-curve by operating at lower densities. The optimum linear velocity for SFC is higher than in HPLC, permitting higher speeds of analyses at about the same efficiency. The speed of analysis for GC is, however, much larger than for SFC on account of the higher interdiffusion coefficients in low pressure versus high pressure gases. With respect to velocity gradients, it may be anticipated that for SFC the efficiency decreases more slowly with increasing velocity than in HPLC if the high velocity branch of the van Deemter curve for SFC proves to be generally of a lower slope.

Using present day equipment for packed column SFC, velocity gradients can often be implemented whereby these gradients are independent of pressure-, temperature-, and composition-gradients which may or may not be run simultaneously to the velocity gradient. Employing present day equipment for capillary SFC, however, this independence is not usually available. Most instruments control column pressure by the volume feed rate of the pump because they possess no variable pressure control device downstream of the column, relying on invariable capillary restrictors instead. However, because recently additional work on variable pressure devices, suitable for capillary columns, has been reported [44, 45], the situation may change in the future.

Acknowledgement. We are grateful to the Arbeitsgemeinschaft Industrieller Forschungsvereinigungen for financial support.

References

1. Hütz A, Schmitz FP, Leyendecker D, Klesper E (1990) J Supercritical Fluids 3:1
2. Schmitz FP, Leyendecker D, Leyendecker D, Klesper E (1989) In: Yoshioka et al. (eds) Progress in HPLC, vol 4, p 3, VSP
3. Hütz A, Leyendecker D, Schmitz FP, Klesper E (1990) J Chromatogr 505:99
4. Leyendecker D, Leyendecker D, Schmitz FP, Klesper E (1987) High Resolut Chromatogr Chromatogr Commun 10:141
5. Leyendecker D, Leyendecker D, Schmitz FP, Klesper E (1986) High Resolut Chromatogr. Chromatogr Commun 9:566
6. Leyendecker D, Leyendecker D, Schmitz FP, Klesper E (1986) J Chromatogr 371:93
7. Leyendecker D, Leyendecker D, Schmitz FP, Klesper E (1987) J Chromatogr 392:101
8. Leyendecker D, Schmitz FP, Leyendecker D, Klesper E (1987) J Chromatogr 393:155
9. Küppers S, Leyendecker D, Schmitz FP, Klesper E (in press) J Chromatogr
10. Klesper E, Schmitz FP (1987) J Chromatogr 402:1
11. Klesper E, Schmitz FP (1988) J Supercritical Fluids 1:45
12. Gemmel B, Schmitz FP, Klesper E (1988) J High Resolut Chromatogr Chromatogr Commun 11:901
13. Gemmel B, Schmitz FP, Klesper E (1988) J Chromatogr 455:17
14. Schmitz FP, Klesper E (1987) J Chromatogr 388:3
15. Klesper E, Schmitz FP (1988) In: White CM (ed) Modern supercritical fluid chromatography. Hüthig, Heidelberg, p 1
16. Leyendecker D, Leyendecker D, Schmitz FP, Klesper E (1987) Chromatographia 23:38
17. Sie ST, Rijnders GWA (1967) Separation Sci 2:755

18. Sie ST, Rijnders GWA (1967) Separation Sci 2:729
19. Wenclawiak BW (1988) Fresenius' Z Anal Chem 330:218
20. Mori S, Saito T, Takeuchi M (1989) J Chromatogr 478:181
21. Bartle KD, Clifford AA, Kithinji JP, Shilstone GF (1988) J Chem Soc Faraday Trans I, 84:4487
22. Berger TA (1989) J Chromatogr 478:311
23. Schomburg G, Roeder W (1989) J High Resolut Chromatogr 12:218
24. Saunders CW, Taylor LT (1989) Chromatographia 28:253
25. Taguchi M, Nagata S, Matsumoto K, Tsuge S, Hirata Y, Toda S (1988) In: Perrut M (ed) Proceedings of the international symposium on supercritical fluids, vol 1. Nice, Oct 17–19, 1988, p 485
26. Anton K, Periclés N, Fields SM, Widmer HM (1988) Chromatographia 26:224
27. Giorgetti A, Periclés NP, Widmer HM, Anton K, Dätwyler P (1989) J Chromatographic Sci 27:318
28. Leyendecker D (1988) In: Smith RM (ed) Supercritical fluid chromatography. Royal Society of Chemistry, Chromatography Monographs, Letchworth, p 53
29. Hirata Y (1984) J Chromatogr 315:39
30. Raynor MW, Davies IL, Bartle KD, Clifford AA, Williams A, Chalmers JM, Cook BW (1988) In: Creaser CS, Davies AMC (eds) Proceedings of an international conference "Across the spectrum: anal appl of spectroscopy" incorporating the first international new infrared spectroscopy 12–15 July 1987, Norwich, UK. The Royal Society of Chemistry, p 227
31. Hirata Y, Nakata F (1986) Chromatographia 21:627
32. Hirata Y, Nakata F, Kawasaki MJ (1986) J High Resolut Chromatogr Chromatogr Commun 9:633
33. Wright BW, Smith RD (1985) Proceedings of the international symposium on capillary chromatography, Riva del Garda, p 910
34. Crow JA, Foley JP (1989) J High Resolut Chromatogr 12:467
35. Strubinger JR, Parcher JF (1989) Anal Chem 61:951
36. Selim MI, Strubinger JR (1988) Fresenius' Z Anal Chem 330:246
37. Strubinger JR, Selim MI (1988) J Chromatogr Sci 26:579
38. Sprinston SR, David P, Steger J, Novotny M (1986) Anal Chem 58:997
39. Steuer W, Schindler M, Erni F (1988) J Chromatogr 454:253
40. King JW (1989) J Chromatographic Science 27:355
41. Saito M, Yamauchi Y, Kashiwazaki H, Sugawara M (1988) Chromatographia 25:801
42. Küppers S, Lorenschat B, Schmitz FP, Klesper E (1989) J Chromatogr 475:85
43. Hirata Y, Nakata F, Kawasaki M (1986) Proceedings of the 7. international symposium on capillary chromatography in Gifu, p 598
44. Raynie DE, Markides KE, Lee ML, Goates SR (1989) Anal Chem 61:1178
45. Berger TA, Toney C (1989) J Chromatogr 465:157
46. Smith RD, Fulton JL, Petersen RC, Kopriva AJ, Wright BW (1986) Anal Chem 58:2057
47. Huston CK Jr, Bernhard RA (1989) J Chromatogr Sci 27:231
48. Berger TA (1989) J High Resolut Chromatogr 12:96
49. Berger TA (1989) Anal Chem 61:356
50. Reid RC, Prausnitz JM, Poling BE (1986) The properties of gases and liquids, 4th edn. McGraw-Hill, p 29
51. Ndiomu DP, Simpson CF (1989) Analytical Proceedings 26:65
52. Pinkston JD, Bowling DJ, Delaney TE (1989) J Chromatogr 474:97
53. Geissler PR (1989) JAOCS 66:685
54. Schmitz FP, Gemmel B, Leyendecker D, Leyendecker D (1988) J High Resolut Chromatogr Chromatogr Commun 11:339
55. Raynor MW, Davies IL, Bartle KD, Clifford AA, Williams A, Chalmers JM, Cook BW (1988) J High Resolution Chromatogr Chromatogr Commun 11:766
56. Fields SM, Grether HJ, Grolimund K (1989) J Chromatogr 472:175
57. Knowles DE, Richter BE, Andersen MR, Later DW (1989) Chimicaoggi, Jan/Feb, p 11

58. Grolimund K, Jackson WP, Joppich M, Nussbaum W, Anton K, Widmer HM (1986) Proceedings on the 7th international symposium on capillary chromatography, Gifu, p 625
59. Davies IL, Xu B, Markides KE, Bartle KD, Lee ML (1989) J Microcolumn Separations 1:71
60. Lurie IS (1988) LC-GC 6:1066
61. Parcher JF, Strubinger JR (1989) J Chromatogr 479:251
62. Martire DE (1989) J Chromatogr 461:165
63. Jacobsen RT, Stewart RJ (1973) J Phys Chem Ref Data 2:757
64. Bartle KD, Boddington T, Clifford AA, Shilstone GF (1989) J Chromatogr 471:347
65. Schoenmakers PJ, Verhoeven FCCJG (1986) J Chromatogr 352:315
66. Roth M, Ansorgova A (1989) J Chromatogr 465:169
67. Simpson RC, Gant JR, Brown PR (1986) J Chromatogr 371:109
68. Berry AJ, Games DE, Perkins JR (1986) J Chromatogr 363:147
69. Mourier PA, Caude MH, Rosset RH (1987) Chromatographia 23:21
70. Leyendecker D, Schmitz FP, Leyendecker D, Klesper E (1985) J Chromatogr 321:273
71. Gere DR, Board R, McManigill D (1982) Anal Chem 54:736
72. Sie ST, Rijnders GWA (1967) Separation Sci 2:699
73. Engelhardt H, Gross A, Mertens R, Petersen M (1989) J Chromatogr 477:169

6 Injection Techniques in SFC

TYGE GREIBROKK

6.1 Introduction

The objective of the injector is to transfer the whole sample or a predetermined fraction of the sample quantitatively to the column in the narrowest possible band. The first part of this objective, a quantitative transfer of all components, is not always attained by every injection method. Sample discrimination effects which are well known from gas chromatography can be found in SFC as well, but this does not necessarily constitute a problem, as long as the user is aware of the possible appearance of such phenomena. The second part of the objective, transfer in a narrow band, is a process which is affected by numerous factors, such as diffusion, linear velocity, temperature, solvents, loading and mixing efficiency. Some of these factors also influence the first part of the objective, the yield of the transfer process.

6.2 The Physical State of the Sample

The physical state of the sample describes whether the sample is in a solid, liquid, gaseous or a supercritical matrix:

- Gaseous samples are usually not introduced directly in SFC since large volumes are needed and since gas chromatography often would be a more appropriate technique.
- Solid samples can be extracted by the mobile phase in a thermostated extractor and the sample is introduced in the supercritical state through a valve.
- Samples dissolved in a liquid can be injected by different methods with inherently different properties.

6.3 Introduction of Supercritical Fluid Extracts

The method for transferring an aliquote of the extract or the complete extract depends on the extraction procedure. Both static and dynamic extractions are utilized. In dynamic extractions the fluid is pumped continuously through the extractor to the collector. If a static extraction is performed, an aliquote is sampled by a valve at intervals. Since static procedures are time consuming and not very efficient, dynamic extractions are currently used more often. In analytical procedures dynamic extraction demands the presence of a sample concentrating unit, based on cooling (cold trap, cryo focusing), heating (density focusing) in combination with a precolumn, or solute precipitation in a piece of tubing by venting towards the atmosphere. Depending mostly on the volatility of the sample constituents, the solute collection will be more or less quantitative.

By extraction with fluids like carbon dioxide, the use of organic solvents which often create environmental problems and need to be disposed of later, can be avoided. Another advantage is that if the mobile phase is used for extraction, the whole extract can be expected to elute through the column. As for other chromatographic systems, there is never a guarantee that all the constituents of a sample dissolved in a solvent will be eluted, depending on the solubility in the mobile phase. However, one should not forget that if the extraction has been performed at high density, elution of all components is expected to require a density gradient with a high density end. The only theoretical possibility of losing part of the sample on the column then is by using packed columns with higher adsorptive properties than the original sample matrix.

Thus, the use of supercritical fluid extraction (SFE) in sample introduction is expected to increase compared to other methods. Still, most samples will probably remain to be found as solutions in organic solvents.

6.4 Introducing a Solution

In most valves currently used in SFC the sample is introduced as a liquid plug in the stream of fluid. With capillary systems the bore of the connections in the valve does usually not exceed 0.25 mm and if the column is directly connected to the injector, the small internal dead volumes result in limited mixing and the major part of the sample arrives at the column inlet dissolved in a liquid composed of the solvent partially mixed with the fluid. Since most injectors are kept at room temperature or only slightly above the critical temperature of the fluid, most samples enter the column in the liquid phase. The transient elution strength of this liquid may be quite different from that of the mobile phase, and can adversely affect the peak shapes.

In a standard capillary system approximately half the injected amount was flushed out of the injector after 0.25 s [1]. An injection time of 1 s was needed for quantitative injection from a 60 nl loop, since reducing the injection time from 5 s to 1 s reduced the amount entering the column with only

3%. With other flow-rates the injection times will of course be different. The last traces of solvent, however, need time to elute from the loop. It is well known that an exponential decay injection profile is obtained by the full-injection method, because of the laminar flow of the displacing mobile phase [2]. By moving the valve back to the load position from the inject position by the "moving injection technique" [3], the solvent tail can be removed.

By injecting very small volumes, by splitting or by using small sample loops, a more thorough mixing and dissolution in the mobile phase can be achieved prior to the column. Today the nominal volume of the smallest commercially available sample loop is 60 nl, but the actual volume may vary by a factor of two [1]. Small volume (<100 nl) sample loops are difficult to produce with high accuracy and the volume is difficult to maintain without changes, particularly with loops made from polymeric materials. Another disadvantage with extremely small loops is that the probability for surface adsorption in the injector increases with an increasing ratio of surface area to volume.

Even 60 nl injections may prove too much for a capillary column, depending on the solute retention. With a capacity factor of 1, the injected volume that produce 10% peak broadening was calculated by Lee et al. [4] to 24 µl on a 50 µm i.d. column, and with a capacity factor of 10 the volume was calculated to 130 µl. Column overloading is rarely caused by the mass of the solutes, but almost always by the amount of solvent in the sample.

The main disadvantage of being forced to introduce small volumes is the requirement for concentrated sample solutions. This removes the possibility to do trace analysis and the highly concentrated solutions also increase the memory effects in subsequent injections.

Thus, injection methods which include solvent removal and peak focusing allow larger sample loops to be utilized with more accurate volumes, the memory effects are reduced, the potential for trace analysis is improved and the quantitation may often be improved as well.

6.5 Peak Focusing

If a 1 µl sample volume is injected into a 50 µm i.d. column, 0.5 m of the column is filled by the solvent plug alone. Consequently peak focusing of volumes larger than this on short and narrow uncoated retention gaps is not feasible in SFC. The main function of retention gaps in SFC is to aid the mixing process of the solvent with the fluid, in order to reduce the elution strength of the sample band, promoting peak focusing at the column inlet. With a lowered elution strength the solutes in the front of the band slow down as they enter the column, allowing the rear part to catch up.

Precipitation of the solutes on the retention gap by evaporation of the solvent is another way of introducing the sample, but this requires a larger retention gap, inert gas flushing and venting towards the atmosphere, as demonstrated by Davies et al. [5]. By using a coated precolumn, with a smaller film thickness than on the main column, a *phase ratio focusing* can be obtained, as shown by Greibrokk et al. [6, 7]. With a less interacting stationary phase on the precolumn *selectivity focusing* or a combination of phase ratio and selectivity focusing can be obtained [7]. With different temperatures on the precolumn/column *density focusing* can be achieved [5, 6, 8].

6.6 Direct Injection

Direct injection is the standard method of sample introduction in packed column SFC. In a high-pressure LC injector, with internal or external sample loop, the sample loop is placed in the fluid path and the sample is swept onto the column by the fluid. Temperatures exceeding room temperature is usually only used with solutes which need higher temperatures to stay in solution.

With open tubular columns and with narrow packed columns direct injection has obvious disadvantages. If the injection is made at low density in order to obtain peak focusing at the column inlet, and the solubility of the solutes is critical, part of the sample may precipitate in the injector, resulting in bad accuracy and ghost peaks by subsequent injections. Low density injection also results in broad solvent peaks which may last for more than 10 minutes, depending on the column length. Injections at higher density or higher temperature result in a more narrow solvent peak, but high density injections can have a disastrous effect on the peak shape of the peaks eluting close to the solvent [9]. Injection at high temperature should be performed with great care, since a solvent which starts to evaporate when the sample is injected in a hot injector results in sample loss and poor reproducibility.

6.7 Open Split (Dynamic Split) Injection

Open split injection has been the most widely used method for introducing the sample on the capillary column, at least until recently. The injection is usually performed at room temperature using a 60 nl or 200 nl sample loop in an internal loop LC injector (Fig. 1). In the splitter, which is a refinement of the concentric tube first described by Ettre and Averill [10], part of the

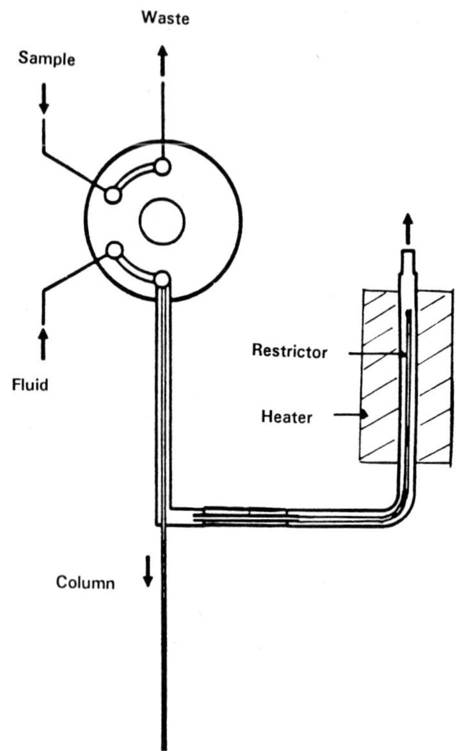

Fig. 1. Open split (dynamic split) injector

sample enters the column while the larger part of the sample moves outside the column until it is vented through a restrictor. Care must be exercised to position the column properly in the splitter assembly. Considerable peak area variations, depending on column positions, was reported by Richter et al. [11]. By changing the dimensions of the split assembly, peak area variations of up to two orders of magnitude was shown by Köhler et al. [12]. Both results are directly connected to the relationship between diffusion and velocity.

As in gas chromatography there is no linear relationship between peak area and split ratio, and the split ratio changes with sample viscosity and mobile phase density. If injections are made at different densities, good quantitation is difficult to obtain.

Most split systems function better with a high split ratio, i.e. with a high proportion of the sample going to waste [12]. A high linear velocity of the fluid passing on the outside of the column is expected to improve the homogeneity of the sample. A high linear flow in the restrictor will also reduce the probability for accumulation of precipitated solutes in the restrictor which leads to changes in the split ratio.

Based on raw peak areas relative standard deviations of 2%–8% have been obtained by open split injection [12]. With internal standards this improved to 1% or better. With "difficult" solutes and certain restrictors a lower performance can be expected.

6.8 Timed Split Injection

With precise computer timing in conjunction with high-speed pneumatics to move the injection valve from load to inject and back to load, the sample is split into one portion which is placed directly on the column and another portion which remains in the sample loop. In the next loop-filling the remaining sample is removed. Less sample discrimination can be expected to be obtained with timed split injection, compared to open split injection.

The relationship between peak area and injection time can not be expected to be linear and consequently non-linear time curves are usually found (Fig. 2). If the timed split injector at the same time is equipped with a small open split, as on some instruments, the exact amount which is introduced on the column is not easily obtainable. In general, quantitation by the internal standard method is recommended for this injection technique, as for all methods based on splitting.

The fraction of the sample which enters the column depends on the viscosity of the sample. With solvents of higher viscosity smaller volumes are delivered to the mobile phase during the period in which the sample loop is connected to the mobile phase flow [13].

Timed split injection is usually considered to result in better reproducibility than open split injection. With internal standards Richter et al. [11] ob-

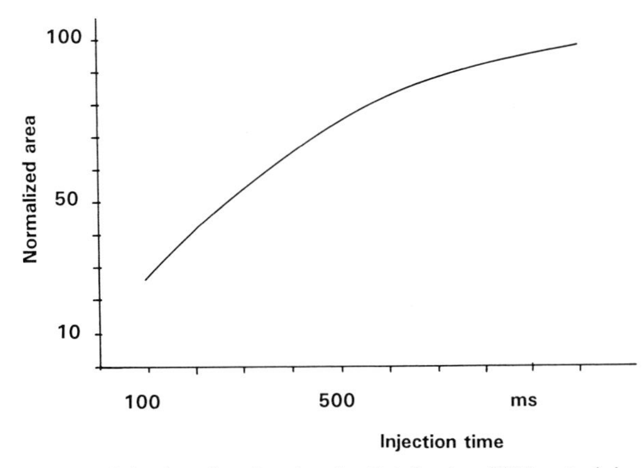

Fig. 2. Peak area as a function of the injection time for timed split injection. With a 1 s injection 9 ng of tocopherol was introduced on the column, while 100 ms injection introduced 2.5 ng

tained relative standard deviations of better than 1% with injection of *n*-alkanes, and similar results have been obtained with the injection of tocopherol [1]. Schomburg et al. [13], however, found little difference in reproducibility between timed split and open split injection of fatty acid methyl esters.

A major advantage of timed split injection is the removal of the solvent tail. The method also allows larger, more accurate, sample loops to be used with capillary columns. Timed split injectors are currently available on most commercial instruments.

6.9 Solvent Effects on Peak Shape

Sample solvents may act as weaker solvents compared to the supercritical fluid [14], but the usual effect, particularly in supercritical carbon dioxide, is that the solvent appears to have a higher solvent strength than the fluid. The bandbroadening effect of sample solvents, which is well known from LC, is also found in SFC [15] and split peaks are another effect of the injection of excessive volumes [9, 11, 16]. An example of peak distortions caused by the sample solvent is shown in Fig. 3c and d.

Although peak splitting and peak shoulders may be caused by various phenomena in the injector or at the column inlet, as known from GC and LC, an important mechanism in SFC seem to be incomplete mixing of the injection solvent with the fluid. Part of the sample elutes with the solvent plug until complete mixing is obtained on the column, while another part of the sample is slightly retained by partitioning on the first part of the col-

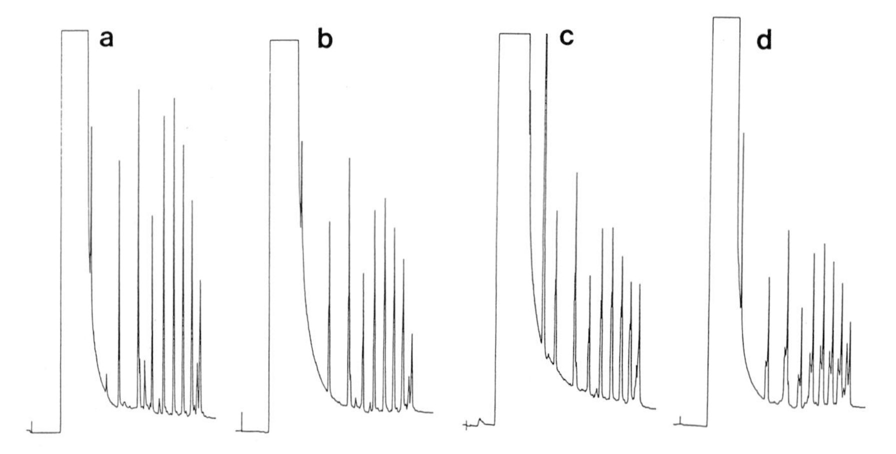

Fig. 3a–d. Peak shapes of triglycerides dissoved in ethyl ether (**a**), acetone (**b**), tetrahydrofuran (**c**) and 2-propanol (**d**). From Ref. [17] with permission of Aster Publishing Corporation

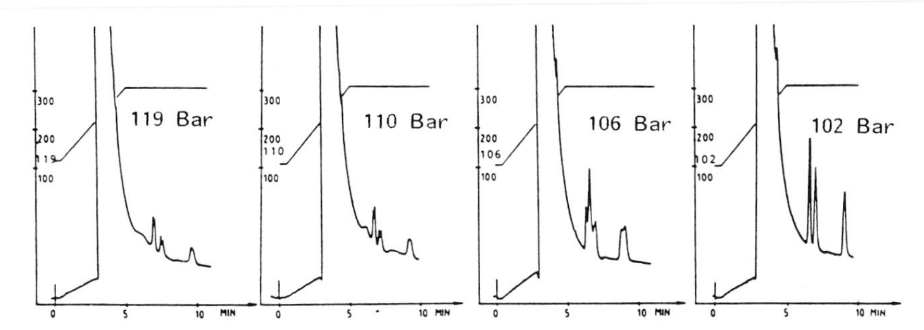

Fig. 4. Direct injection of palmitic acid, cholesterol, tripalmitine and cholesteryl palmitate in 200 nl chloroform on a 50 μm i.d. column at various pressures. From Ref. [9] with permission from Dr. Alfred Huethig Verlag

umn, as suggested by Berg and Greibrokk [9] and by Hirata et al. [16]. This is a process that has been shown to take place on the column/precolumn, since the splitting disappeared by raising the column temperature, apparently due to the increased diffusion at higher temperature [9].

The initial pressure may also have an effect on the peak shape. At lower pressure the increased diffusion rate and the increased diffusion time result in better mixing and peak splitting disappears (Fig. 4).

By including small mixing chambers between the injection valve and the column, Hirata and Inomata [17] showed that the peak splitting which otherwise resulted from 0.2–1 μl injections on 100 μm i.d. open tubular columns, could be removed. Improved mixing is probably also the reason why an uncoated retention gap was reported to result in less discrimination [4].

6.10 Solvent Venting with a Precolumn

In the first solvent venting experiments, sample volumes of up to 0.5 μl were injected on a 50 μm i.d. precolumn with complete solvent elimination [6, 7]. With this technique, the solvent is more or less completely separated from the solutes on the precolumn by a partitioning mechanism, and the solutes are transferred to the main column and focused at the column inlet (Fig. 5). The major part of the solvent is vented through a restrictor at a rapid (10–20 cm/s) but controlled flow rate. With a 2 m precolumn the venting time is only 15–20 s. Venting times which are too long discriminate the early eluting peaks (Fig. 6) and too high starting pressure degrades the peak shape. The peak focusing at the column inlet is based on phase ratios, by using a thicker film on the column than on the precolumn, or by partition coefficients, by using a more retaining stationary phase on the column, or

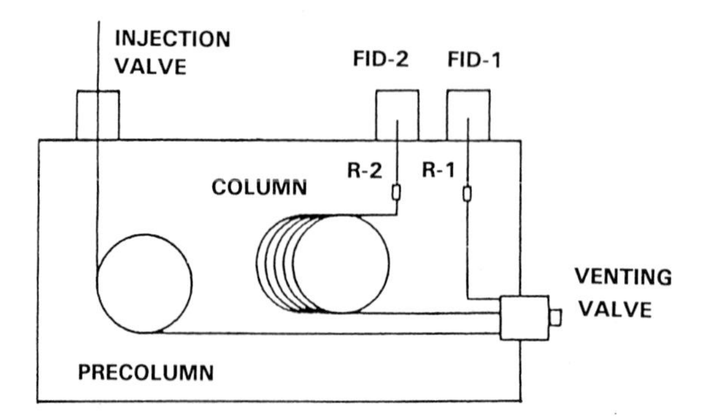

Fig. 5. Instrumentation for solvent venting with a precolumn. From Ref. [9] with permission of Dr. Alfred Huethig Verlag

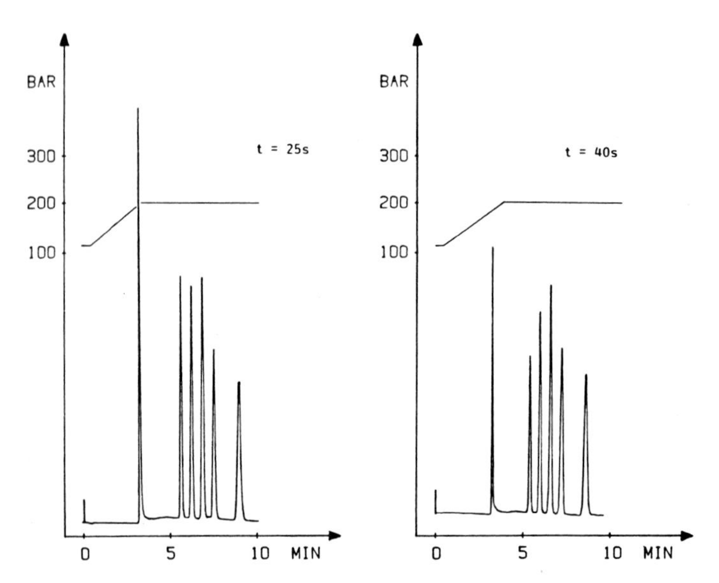

Fig. 6. Discrimination of early eluting peaks caused by an excessively long venting time (40 s vs 25 s). The first peak in the chromatogram is the solvent peak

both. By choosing conditions whereby an equilibrium was attained on the precolumn, the sample recovery of palmitic acid, cholesterol, tripalmitine and cholesteryl palmitate was 100%. With the $2\ m \times 50\ \mu m$ i.d. precolumn, however, 1 μl injections resulted in a substantial loss of the sample (Fig. 7). Thus, longer or wider precolumns are required for injection volumes of 1 μl or more.

Based on absolute area measurements the solvent venting technique resulted in better injection reproducibility (RSD = 2% – 5%), compared to timed split injection (RSD = 6%) and open split injection (RSD = 8%). Addition of an internal standard reduced the relative standard deviation to less than 1%.

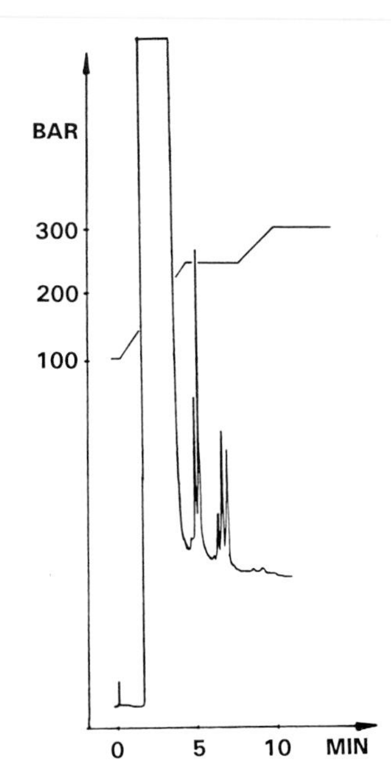

Fig. 7. 1 µl injection of glycerides in chloroform on a 50 µm i.d. column with solvent venting. The small precolumn (2 m×50 µm) is overloaded, as seen by the large solvent peak, and part of the sample is lost with the solvent. With injections of 1 µl or more, longer precolumns are required

This method of solvent elimination, which is based on a rough separation of solvent and solutes by partitioning on a precolumn, seems likely to be a method for sample volumes not larger than 1–5 µl. In this range concentrations as low as 0.1 ppm can be analyzed directly, with flame ionization detection. The valve between the precolumn and the column produce some band broadening, which must be taken care of by peak focusing. The valve gives, however, additional opportunities, i.e. to use the system for coupled column separations.

6.11 Solvent Backflush

Another method for complete elimination of the sample solvent is the backflushing technique developed by Lee et al. [4]. The instrumentation is based on the delayed split injection technique which includes an on/off valve in the vent line (Fig. 8). The sample is injected with the split vent valve closed (off). After a time delay of approximately 1 min, the injection valve is closed and the split valve is opened simultaneous with a rapid negative pressure ramp. The rapid depressurization causes a density gradient to propagate along the column from the inlet, and a reversed flow develops in the first part of the column. The solvent backflushes through the open vent line and the solutes precipitate on the column wall. By reversing to a positive density program, the chromatogram is developed (Fig. 9b). A similar meth-

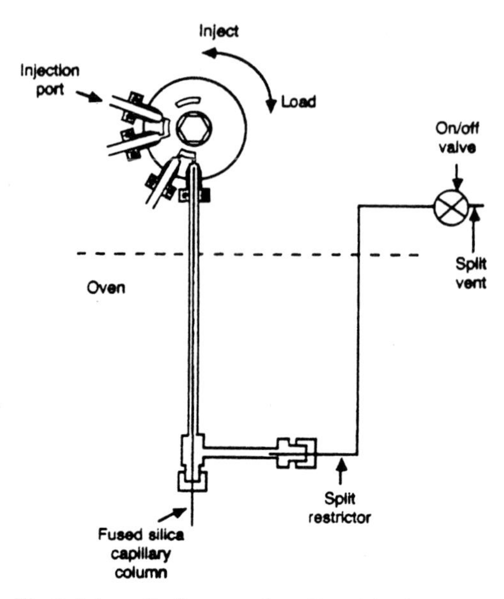

Fig. 8. Schematic diagram of a splitter injection system that can be used for open split, delayed split or solvent backflush sample introduction. From Ref. [4] with permission of Aster Publishing Corporation

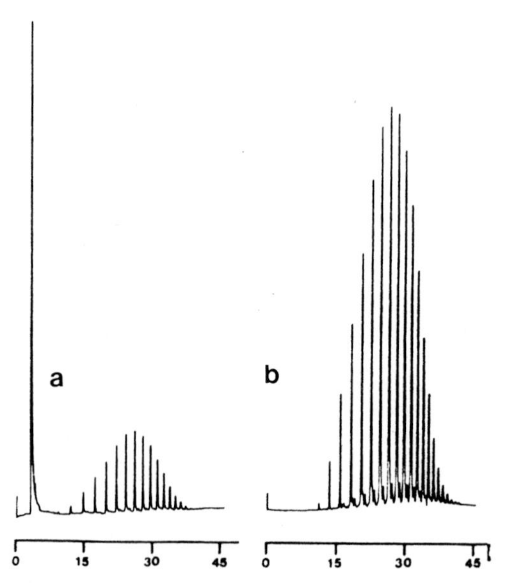

Fig. 9 a, b. Delayed split injection **(a)** and solvent backflush injection **(b)** of Triton X-100 using delay times of 1 s and 15 s. Only 20% of the sample was transferred to the column by the delayed split injection. From Ref. [4] with permission of Aster Publishing Corporation

od, but without the negative pressure ramping, was called delayed split injection (Fig. 9a). Preliminary measurements from the injection of 0.2 μl volumes with the solvent backflush technique showed a recovery of 68% of naphthalene and 85% of coronene. The lower percentage for naphthalene is due to its higher vapor pressure and greater loss with the solvent during backflushing.

Another backflushing method, called postinjection solvent venting by Hawthorne and Miller [18], making use of a short (7 cm), wide (0.3 mm) retention gap and syringe injection allowed injection of 0.5 μl volumes on the depressurized column, with detection limits of 0.2 ppm of alkanes. Negative pressure ramping was not included, possibly due to the wide injection port and retention gap which facilitates rapid depressurization. The backflushing started 15 s after the pressure program was initiated and lasted for 5 s. The peak area reproducibility was approximately 10% on raw areas and 0.9% – 2.5% with an internal standard. The appearance of several solvent peaks from the same solvent could be explained partly by the pressurization/depressurization processes [16]. Solvent peaks or humps, which affect the quality of the first part of the chromatograms, could be removed by increasing the injector temperature from 45° to 65 °C [18].

Hirata et al. [16] used a mixing chamber and a delayed purging of the mixing chamber, called purged splitless injection. The relative standard deviations were 6.4% on peak heights and 1.2% with internal standards.

6.12 Solvent Venting with Gas Purging

Another solvent elimination technique, based on gas purging, has been developed by Lee et al. [4]. By purging the sample into a precolumn with an inert gas, i.e. by GC conditions, the solvent is evaporated in the gas flow and carried through the column and out of the vent valve, which is located between the column and the detector (Fig. 10). The precolumn can simply be the first part of the column which extends out of the oven. After the solvent is evaporated and the solutes are precipitated on the walls of the precolumn, the purging valve is switched back to the carbon dioxide line, and liquid CO_2 is introduced. The solutes which dissolve in the flow are than focused by the temperature gradient at the column entrance.

The reproducibility of 0.2 μl injections was found to be better than 0.9% (with internal standard). Alkanes could be analyzed in 0.5 ppm concentrations with 0.2 μl injections. With a 2.5 m precolumn, 0.5 μl volumes could be injected and purged before developing the chromatogram. Compounds that are soluble in the solvent, but less soluble in carbon dioxide, may be lost, at least partially, by remaining adsorbed to the precolumn walls. Polar compounds may also lag or adsorb to the walls of the sample loop, a problem which can be solved by injecting the sample directly into the precolumn

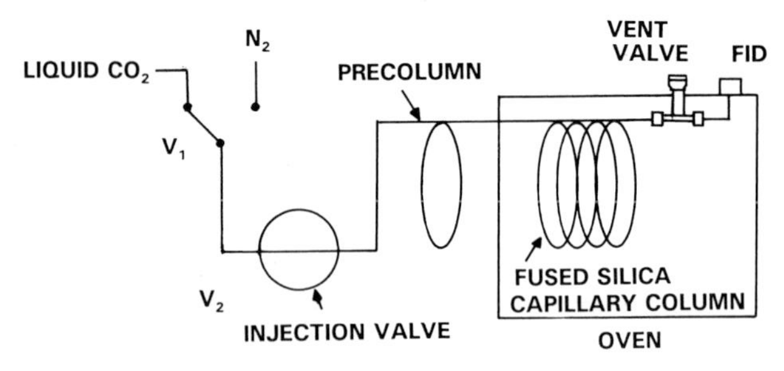

Fig. 10. Schematic diagram of gas purging-solvent venting injection. From Ref. [4] with permission of Aster Publishing Corporation

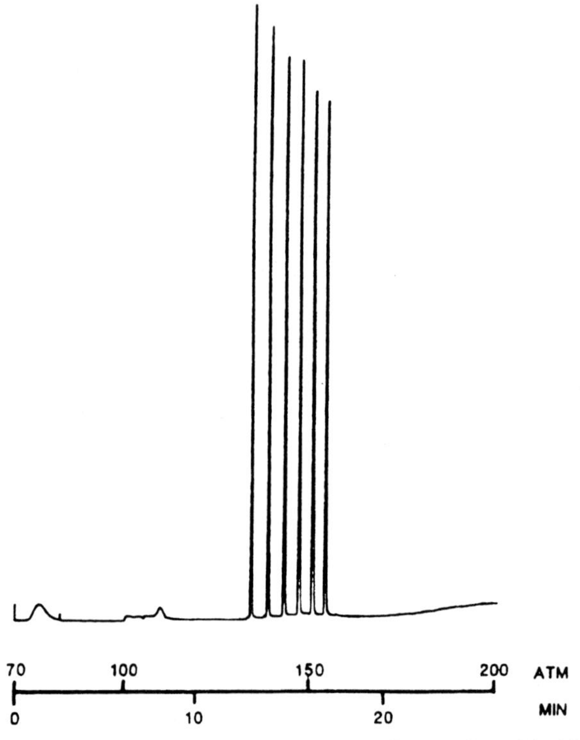

Fig. 11. Gas purging-solvent venting of *n*-alkanes (50 ppm) in 0.2 μl hexane on a 50 μm i.d. column. From Ref. [4] with permission of Aster Publishing Corporation

with a syringe. Due to the resistance of the column towards the purging gas, this method is not quite as rapid as the one with the vent valve between the columns With a 2.5 m precolumn and a 4 m column, 0.2 μl of a 50 ppm (each) solution of alkanes in hexane gave the chromatogram in Fig. 11. The solvent elimination time was 3 min. with nitrogen at 50 bar. If the temperature gradient is not sufficient for focusing the solutes, partitioning or phase ratio focusing may become necessary.

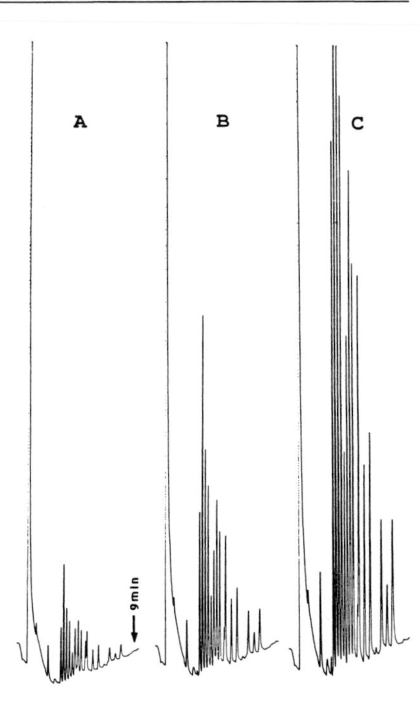

Fig. 12 A – C. Gas purging-solvent venting of 1, 6 and 10 μl samples of *n*-alkanes in carbon disulfide on a 1 mm i.d. packed column. From Ref. [19] with permission of Dr. Alfred Huethig Verlag

The major efforts so far in developing methods for solvent elimination have been in connection with the use of capillary columns, but recently a study of similar techniques for packed microbore (1 mm i.d.) columns was performed by Dean and Poole [19]. By gas purging the sample into a steel retention gap, 1 – 10 μl injections of *n*-alkanes in 0.1 – 1 ppm concentrations resulted in high peaks with good peak shapes (Fig. 12). The reproducibility of 1 μl injections of 1 – 10 ppm concentrations was 0.7%, by the internal standard method.

Both the solvent backflush methods and the gas purging-solvent venting methods are most effective if the solvent is much more volatile than the solutes of interest, otherwise loss of the more volatile components in the sample is possible. The solvent venting method based on partitioning on a precolumn can lead to loss of components with capacity factors close to the solvent, and since these components almost always are volatile, all the solvent elimination methods suffer from essentially the same weakness. For the majority of applications in SFC this weakness is not a serious problem, however, since the usual components of interest are compounds of relatively low volatility.

In conclusion, sample volumes of approximately 1 μl can be injected on 50 μm i.d. columns with current solvent elimination methods. If there is a need to inject considerably larger volumes, multiple injections on retention gaps/precolumns or sample concentration on packed precolumns are required. Whether packed precolumns can be utilized is completely dependent on the properties of the sample.

6.13 Sample Losses in the Injector

6.13.1 Leaks

With pressures possibly approaching 500 bar, occasional leaks are bound to occur in a rotating valve. Many valves are guaranteed leak-proof to only 400 bar, but even below this limit leaks arise. Leaks are often a result of the impact of solid particles between the rotor and the stator, particles coming from the sample, the fluid tank or from the end of the connecting tubing. Particle filters should be inserted in the line prior to the injector. Fused silica tubing, both at the inlet and the outlet, should be connected with care to avoid splintered ends. Samples with suspended particles need to be filtered.

Leaks are, however, not so easily detected since the gas flow which may escape is very low. The reduced peak-height, which is the result of a leak in the valve could also have other causes, such as detector problems. A close inspection of the stator and the rotor usually will reveal whether there is a leak.

6.13.2 Losses with Volatile Solvents

By injecting the sample dissolved in a volatile solvent in a hot injector, part of the sample will be lost. Even when the injector is not intentionly heated, waste oven heat may warm the valve to a temperature where the expanding gas from evaporating diluent can expel part of the liquid sample, depending on the valve port assignment, as demonstrated by Chester and Innis [20]. A waste-port restrictor [20] or a valve cooling system can solve the problem.

6.13.3 Discrimination

If the yield of the transfer to the column is unequal for the solutes, the term discrimination is commonly used. Discrimination may take place when a part of the sample is split away, as with open split injection, solvent venting, solvent backflush and similar methods. The more volatile/rapidly eluted components are transferred to the column to a smaller extent than the less volatile/more retained components. Claims of no discrimination with split systems should be regarded as sample related statements, and the actual extent of discrimination in every case need to be determined with standard solutions. For the majority of applications in SFC, however, discrimination effects are of little significance.

References

1. Greibrokk T, Berg BE, Hohansen H (1988) In: Perrut M (ed) International symposium on supercritical fluids in Nice, October 1988, p 425
2. Scott RPW, Simpson CF (1982) J Chromatogr Sci 20:62
3. Harvey MC, Stearns SD (1983) J Chromatogr Sci 21:473
4. Lee ML, Xu B, Juang EC, Djordjevic NM, Chang H-CK, Markides KE (1989) J Microbiol Sep 1:7
5. Davies IL, Xu B, Markides KE, Bartle KD, Lee ML (1989) J Microcol Sep 1:71
6. Greibrokk T, Berg BE, Blilie AL, Doehl J, Farbrot A, Lundanes E (1987) J Chromatogr 394:429
7. Farbrot-Buskhe A, Berg BE, Gyllenhaal O, Greibrokk T (1988) HRC & CC 11:16
8. Schomburg G, Roeder W (1989) HRC & CC 12:218
9. Berg BE, Greibrokk T (1989) HRC & CC 12:322
10. Ettre LS, Averill WS (1961) Anal Chem 33:680
11. Richter BE, Knowles DE, Anderson MR, Porter NL, Campbell ER, Later DW (1988) HRC & CC 11:29
12. Köhler J, Rose A, Schomburg G (1988) HRC & CC 11:191
13. Schomburg G, Behlau H, Häusig U, Goening B, Roeder W (1989) HRC & CC 12:142
14. Hirata Y, Nakata F (1984) J Chromatogr 295:315
15. Blilie AL, Greibrokk T (1985) Anal Chem 57:2239
16. Hirata Y, Tanaka M, Inomata K (1989) J Chromatogr Sci 27:395
17. Hirata Y, Inomata K (1989) J Microcol Sep 1:242
18. Hawthorne SB, Miller DJ (1989) J Chromatogr Sci 27:197
19. Dean TA, Poole CF (1989) HRC-CC 12:773
20. Chester TL, Innis DP (1989) J Microcol Sep 1:230

7 Stationary Phases for Packed Column Supercritical Fluid Chromatography

COLIN F. POOLE, JOHN W. OUDSEMA, THOMAS A. DEAN, and SALWA K. POOLE

7.1 Introduction

Virtually all contemporary research using packed columns in SFC is performed with column packings and column dimensions optimized for high pressure liquid chromatography (HPLC) [1]. This catalyzed the early development of practical applications of packed column SFC and provided a variety of stationary phase chemistries to optimize individual separations with the limited selection of available supercritical fluids with favorable critical constants. The lack of a reliable theoretical framework for kinetic optimization of column parameters has also served to deflect attention from the question of whether columns designed for the separation of low-molecular-weight solutes with incompressible liquids are necessarily ideal for the typical separations most suited to SFC [2]. It is still too early to answer this question definitively but at least some positive statements can be made concerning column dimensions, packings, and operating conditions that are most useful for SFC.

7.2 Physical Properties of Column Packings

Quite a wide range of column packings have been developed for HPLC and most of these have also been used to various extents in SFC. Totally porous silica, alumina, and pellicular silica packings are used in adsorption chromatography of non polar and moderately polar solutes. The elutropic strength of common supercritical fluids is not very high which prevents the elution of solutes with acidic, basic, or strongly hydrogen bonding functional groups. For this reason chemically-bonded silica-based packings have been the most widely used packings for separating polar solutes. These can be prepared in several ways to give packings with different properties. The most common packings are of the monomeric type prepared by the controlled reaction of monochloro-, monoalkoxyl-, or dialkylaminoalkylsilanes with silanol groups on the silica surface to form$\equiv$Si$-$O$-$Si$-$R groups. The R group is typically methyl, octyl, octadecyl, phenyl, 3-aminopropyl, or 3-cyanopropyl. With polyfunctional silanes one or two bonds with the surface are possible per reagent molecule, Fig. 1 [1]. If trace quantities of water

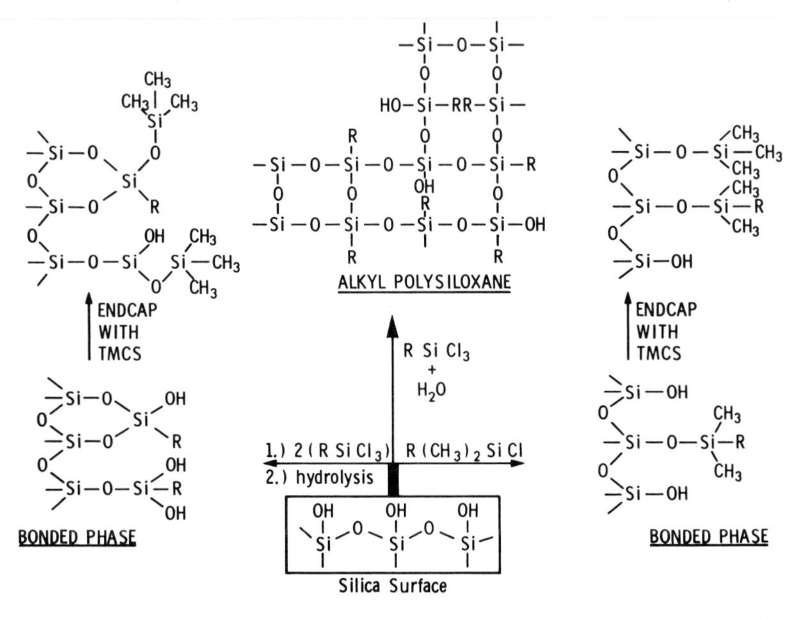

Fig. 1. Preparation of siloxane bonded phases by surface silanization of macroporous silica (from Ref. [1])

are present during the reaction with polyfunctional silanes, polymerization reactions may occur, resulting in phases with a heavier loading of organic material and chromatographic properties that are different from the monomeric phases. The preparation of chemically bonded phases is reviewed in Ref. [1, 3, 4].

The surface of silica consists of various kinds of silanols and siloxane bonds, Fig. 2. The silanols can exist on the surface in single, geminal, or vicinal forms. A fully hydroxylated silica surface contains about 8 $\mu mol/m^2$ of silanol groups [5]. This may be higher than certain grades of chromatographic silicas that are stated to have a silanol concentration in the range 5 – 6 $\mu mol/m^2$. A fully hydroxylated surface is advantageous for preparing chemically bonded phases and is characterized by: (1) a large number of associated silanols; (2) a higher pH; (3) a markedly lower adsorptivity for basic solutes; (4) a significantly improved hydrolytic stability of bonded phase ligands; and (5) increased mechanical stability. Chromatographic grades of silica contain about 0.1% – 0.3% of the silica mass as a wide variety of metal salts. These metals may act as sites for strong solute interactions by chelation. It is also recognized that a small population of the silanol groups on the silica surface interact much more strongly than other, "normal" silanols. These specific active sites interact with solutes about 50 times more strongly than normal silanols [4, 6]. It has been suggested that the metal impurities hidden under the silica surface influence the adjacent silanols increasing their ability to interact, or their strength of

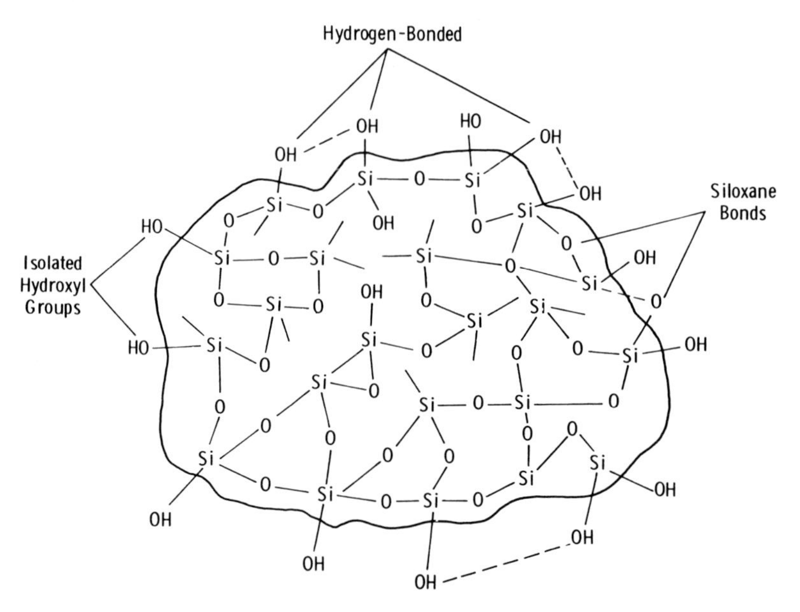

Fig. 2. Structure of silica gel, depicting the various types of bonds and silanol groups present at the surface (from Ref. [1])

interaction, and are thus responsible for the small concentration of strongly interacting silanols on the silica surface. The heterogeneity of the silica surface and the presence of adsorbed impurities, particularly water, are primary sources of the poor reproducibility of both the silica packings and the chemically bonded materials derived from them.

For steric reasons it is impossible to derivatize all silanol groups on the silica surface using mono- or polyfunctional silane reagents [3, 4, 7, 8]. For octadecyldimethylsilyl ligands only about half of the silanol groups may be reacted increasing to about three-quarters for trimethylsilyl ligands. Typical bonded phase loadings correspond to about 3.0–3.9 $\mu mol/m^2$ for octadecyldimethylsilyl ligands and about 2.25–3.50 $\mu mol/m^2$ for dimethylalkylsilyl ligands containing a terminal functional group on the alkyl chain. The curvature of the pores can influence the final coverage of the bonded phase. This effect is of greater importance the longer the alkyl chain anchored on the pore surface and also depends on the pore size distribution. Micropores with diameters less than 2 nm are particularly undesirable. Ideally, silicas used to prepare chemically bonded phases have pore diameters in the range 6–10 nm. The extended confirmation of an octadecylsilyl ligand is 2.45 nm long, preventing maximum coverage being obtained on silicas with narrow pores. Silicas with narrower pores would probably show a tendency to produce ink-bottle-shaped pores after derivatization as the initial reaction of the cylindrical pores at the edge would prevent other molecules of reagent diffusing inside the pores. Smaller pores would only be partially silanized which is why endcapping procedures are necessary to cover as many silanol groups as possible. Endcapping, a subsequent sequential reaction with a trimethylsilane reagent, Fig. 1, is generally unnecessary for bonded layers at maximum coverage on medium to wide pore silicas since an increase in the bonding density is not observed. Bonded layers produced using polyfunctional silanes can have higher silanol concentrations than the starting silica resulting from the hydrolysis of unreacted chlorosilane groups, etc., that are now part of the bonded layer. For example, using the same silica substrate the concentration of silanol groups was determined to be 3.73 $\mu mol/m^2$ after reaction with a monochlorosilane, 6.74 $\mu mol/m^2$ after reaction with a dichlorosilane, and 11.70 $\mu mol/m^2$ after reaction with a trichlorosilane reagent. Endcapping becomes obligatory in this case if inert bonded layers are to be prepared.

For a silica gel with a given pore diameter, the concentration of bonded silane decreases approximately linearly as the *n*-alkyl chain length increases. The concentration of a given bonded silane increases as the pore size of the silica gel increases. For monomeric phases retention decreases with increasing pore size (and with decreasing surface area) in HPLC, but the elution order and relative retention time of the components remain the same. The separation of small molecules in HPLC is best performed using narrow pore silica substrates with surface areas of 150–400 m^2/g. These are the types of chemically bonded packings used in SFC today, but because of the general role of SFC to separate middle-molecular-weight molecules, it can be argued that wider pore packings would be preferred [2]. Finally, it should be noted that silicas from different sources have widely different apparent surface pH values, range 3.9–9.9 [4]. This may be of no concern for the separation of neutral molecules but correlates well with the difficulties observed in separating acidic and basic substances on different packings. Strong acid/base interactions generally result in poor peak shapes and possibly poor recovery of sample mass. This wide range of apparent surface pH values cannot be attributed to properties of the silanol groups, and must be due to surface impurities or contaminants acquired during the manufacturing process.

In the last few years new bonded phase packings have been introduced with a view to increasing their pH stability outside of the range 2–7.5 and to minimize undesirable interactions with silanol groups. Schomburg and others [9–11] have developed polymer encapsulated stationary phases for this purpose. These phases are prepared by mechanically coating silica or persilanized silica particles with a poly(siloxane) or butadiene prepolymer, for example, which is then immobilized by peroxide, azo-t-butane, or γ-radiation induced chemical crosslinking reactions. Silanization decreases the number of effective silanol groups which are then further shielded by the polymeric film anchored over the surface. In reversed-phase liquid chromatography these packings have shown improved chromatographic properties compared with monomeric chemically bonded phases for the separation of basic solutes. Also, since phase preparation does not depend on the chemistry of the underlying substrate, materials other than silica, for example alumina, can be modified. Alumina-based materials are more pH stable than silica, which has been one reason for the interest and development of polymer encapsulated alumina packings in recent years. Polymer encapsulated packings have a film thickness of about 1 nm to maintain reasonable mass transfer characteristics. These films must be permeable, so presumably analytes can still interact with the particle substrate and, since it is known that polymeric films in open tubular columns swell in supercritical fluid mobile phases, this will most likely also occur for the polymer encapsulated packings [12]. Swelling might improve mass transfer characteristics so that thicker films could be used for packings in SFC, but would not inhibit access of the sample to the underlying substrate. Thus, although one might anticipate a reduction in undesirable solute/substrate interactions using polymer encapsulated packings, it is difficult to envisage how these could be reduced to zero by polymer encapsulation.

Since most of the undesirable interactions associated with chemically bonded packings are thought to be due to the presence of unreacted and unshielded silanol groups alternative substrates have been proposed; notably alumina, zirconia, carbon, and macroporous organic polymers. Alumina and zirconia are still chemically active surfaces and are unlikely to yield a universal solution to the problem of preparing inert column packings for either HPLC or SFC. Porous graphic carbon packings with a unique sponge-like structure of high mechanical stability and adequate surface area (ca. 150 m^2/g) and porosity (ca. 80%) recently became available under the trade name Hypercarb [13, 14]. This material is produced by impregnating a silica gel template with a phenol/hexamine mixture, polymerizing this mixture within the pores of the silica gel, pyrolyzing the resin under nitrogen, dissolving out the silica template, and finally heating the remaining porous carbon to a temperature in excess of 2000 °C. Basic substances do not tail as significantly as they do on chemically bonded, silica-based packings in reversed-phase HPLC, but it is not clear that these packings are completely inert either. Any oxidation of the packing during manufacture would introduce polar functional groups into the otherwise inert graphite matrix which could have undesirable chromatographic properties. Also, in reversed-phase HPLC the porous graphitic carbon packings show very strong retention of solutes compared to chemically bonded silica packings, which may limit their application to high-molecular-weight analytes [14, 15]. Such strong, nonspecific, adsorption properties might be a further disadvantage in SFC, which is often applied to the separation of middle-molecular-weight samples.

There are a fairly large number of macroreticular polymeric packings that have been developed for reversed-phase, ion-exchange, and size-exclusion liquid chromatography [1]. Those materials based on poly(styrene-divinylbenzene) with various degrees of crosslinking are most likely to be useful in SFC. They can be used over a wide pH range, but have demonstrated some unfavorable characteristics when used in reversed-phase HPLC

[16–18]. It was noted that the efficiency and peak asymmetry decreased with increasing retention, particularly for nonpolar solutes. Verzele claims that polystyrene phases are fundamentally different from silica gel phases because of their changing microporosity and ability to act like solvents [18]. The main cause of band broadening he attributes to slow diffusion in the polymer matrix. In particular, he attributes most of the poor chromatographic properties of these materials to the presence of micropores (diameter <2 nm) in the polymer structure. The presence of such pores seems unavoidable due to the structure of the polymeric skeleton for polymeric beads having both high mechanical strength and a very dense structure. The micropores are effective at retaining molecules exceeding the exclusion limit, since as large molecules unfold their ends creep into the small pores, resulting in retention of the entire molecule. Even so called nonporous polymeric packings will contain micropores and can provide much higher retention than predicted from their BET surface areas. Although manufacturers try to optimize the production process for minimum swelling and shrinking in different solvents the permeability of the macroreticular polymeric packings is found to change when different solvents are used [17]. It is known that repeated expansion and compression of polymeric beads, as might occur during density programming in SFC, can lead to mechanical breakdown of the bead structure and loss of chromatographic performance [2, 19].

7.3 Influence of Substrate Morphology on the Properties of Chemically Bonded Phases in SFC

Column packings commonly used in HPLC have large surface areas that may not be required for separations in SFC [2, 20, 21]. Large surface areas generate low phase ratios, and this leads to high retention. Secondly, packings with a large surface area contain a greater number of active sites per unit weight and are more difficult to deactivate. For nonpolar molecules the retention behavior with carbon dioxide as the mobile phase is influenced more by the pore dimensions and pore size distribution than the hydrophobicity of chemically bonded packings [20]. For polar compounds retention will tend to depend mainly on the pore dimensions (surface area) and concentration of accessible silanol groups. In general, small pore diameter packings will be preferred for the separation of nonpolar samples, increasing to larger diameters to elute high-molecular-weight samples. For polar samples, larger pore diameter packings will be preferred adjusted to smaller sizes to improve resolution, when required. Since SFC is used to separate low- and middle-molecular-weight solutes size exclusion effects will not generally be a problem, although very little information is known about the conformation of large molecules in supercritical fluids. However, the pore size distribution, and in particular, the concentation of micropores, is very important as far as retention and peak asymmetry are concerned. A disproportionate amount of the total surface area is contained within the micropores, which for steric reasons, contain a disproportionately low coverage of bonded phase. Even large molecules can interact with micropores if part of their pendant structure penetrates the pore and is subjected to strong interactions causing increased retention of the whole molecule and peak tailing if there is a difference in kinetics between interactions in the micropores and macropores. Greater attention will have to be given to pore size distribution problems for silica-based packings in the future. Some useful data is available from HPLC [22].

7.4 Influence of Surface Heterogeneity on the Properties of Chemically Bonded Phases

For steric reasons the surface of chemically bonded packings is always heterogeneous containing different concentrations of chemically bonded and free silanol groups. To a first approximation it can be assumed that the two types of site contribute independently to retention in packed column SFC with nonpolar supercritical fluids such as carbon dioxide, nitrous oxide, and sulfur hexafluoride [23]. This will result in poor peak shapes if the two sites exhibit different kinetics or if the sorption isotherms are of a different type. Schoenmakers et al. [23] have suggested a model that enables the relative number of bonded and free silanol groups to be estimated from the influence of sample size on retention. In less formal terms it has been established that the interaction of sample proton donor/acceptor and dipolar functional groups with free silanol groups of the column packings causes the characteristic peak tailing and sample adsorption or degradation that occurs in packed column SFC with relatively nonpolar fluids [2, 21, 24–27].

7.4.1 Chemical Interactions with Silanol Groups

It is a well established fact that silanol groups can react chemically with alcohols, amines, and isocyanates, at moderate temperatures, to form bonded ligands [1]. Silanol groups can also induce acid catalyzed transformation of labile molecules. Evans et al. [28] used a stop flow technique to determine the extent of dehydration of phenylpropanols on chemically bonded phases under SFC conditions. Significant dehydration was observed for the tertiary alcohol on an octadecylsilanized silica packing, which was significantly reduced using a 3-aminopropylsilanized silica packing, and did not occur at all on a graphitic carbon packing lacking silanol groups. Hirata [29] has reported that silica packings become irreversibly modified when ethanol-containing mobile phases are used in SFC and Schmitz et al. [30] have shown that silica reacts with 1,4-dioxane under SFC conditions to produce an unidentified bonded organic layer. This dioxane-modified silica phase has some useful chromatographic properties and has been used to separate moderately polar compounds having excessive retention or poor peak shapes on silica. An interesting example of a chemical reaction occurring on silica is shown in Fig. 3 for the separation of cholestane, 5-cholestene, 3,5-cholestadiene and cholesterol, at two different temperatures, on a low-surface-area pellicular silica packing. Adsorption interactions are useful for the separation of the three cholestane hydrocarbons which are difficult to resolve on bonded-phase packings. On the other hand, the peak shape for cholesterol on this packing is substantially broadened and shows tailing on its leading edge. Increasing the column temperature causes an increase in peak tailing and a loss of injected mass.

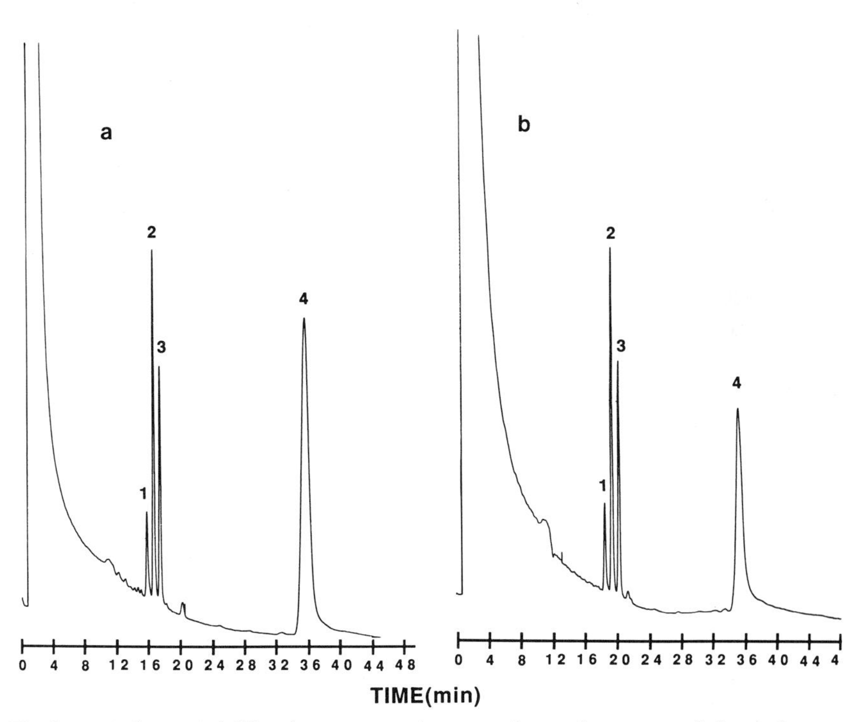

Fig. 3a, b. Influence of Silanol groups on the separation and recovery of *1* = cholestane; *2* = 5-cholestene; *3* = 3,5-cholestadiene; and *4* = cholesterol at **a** = 80 °C and **b** = 100 °C. The column was 10 cm × 1 mm I.D. packed with Corasil (II), a pellicular silica packing. The mobile phase was carbon dioxide and the pressure program 10 min at 800 atm increased linearly to 300 atm over 30 min (from Ref. [2])

Increasing the temperature further results in the complete abstraction of cholesterol from the chromatogram, the condensation of the hydroxyl group of cholesterol with silanol groups of the packing now being complete. This leads to an important distinction between physical and chemical interactions with active surfaces. The extent of a chemical reaction usually increases with temperature while adsorptive interactions, not involving chemical bond formation, usually decrease. Depending on the nature of the retention mechanism it may be feasible to adjust the temperature to an appropriate value to minimize the undesirable interaction while simultaneously adjusting the mobile phase density to provide reasonable retention to separate the sample. In our experience, reactions of a chemical nature persist even on the most chemically deactivated bonded-phase supports commercially available. Low surface area, deactivated, polymer encapsulated phases generally show the lowest activity, but even on these phases, it may be difficult to elute polyfunctional, middle-molecular-weight molecules from the column, or to obtain reasonable peak shapes.

7.4.2 Deactivation of Silanol Groups with Mobile Phase Modifiers

Various quantities of polar modifiers can be added to supercritical fluids of low polarity to increase their solvent strength and to reduce undesirable interactions between unshielded silanol groups and the sample. To improve chromatographic performance modifiers are generally used at concentrations below about 2% (V/V). At these concentration levels the modifier may have only a small influence on the solubility of the sample in the mobile phase and is used primarily to mask interactions between the sample and accessible silanol groups on the stationary phase surface. If these interactions made a major contribution to retention of the sample then in the presence of a suitable modifier a substantial reduction in retention will occur, but more typically, smaller changes in retention with a decrease in peak asymmetry are observed. Modifiers should be considered as a means of extending the deactivation of well-deactivated chemically bonded packings and not as a means of overcoming the deficiency of poorly deactivated packings. The general role of mobile phase selectivity modification by organic solvents and programming techniques using composition gradients will not be considered in this section.

As well as effectiveness as a deactivating agent detector compatibility is an other important consideration for the selection of a modifier. Polar organic solvents cannot be used with the popular flame ionization detector because of their large detector responses. Water [31–33] and formic acid [33, 34] are the most popular modifiers used with the flame ionization detector to improve the chromatographic properties of proton donor or acceptor solutes, generally, and occasionally neutral polar molecules. With UV detection the same modifiers can also be used, as well as polar organic

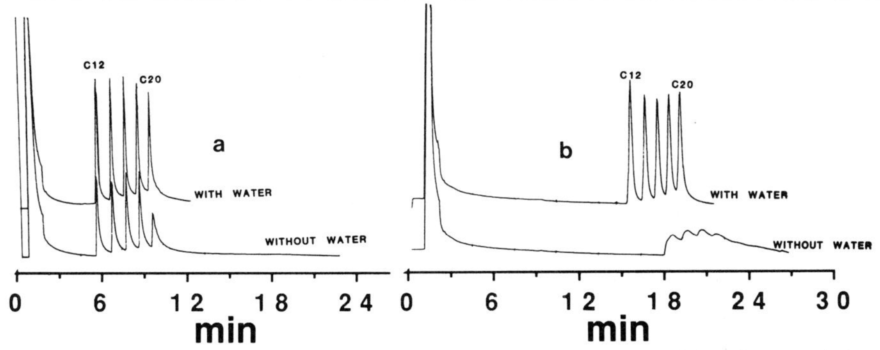

Fig. 4a, b. Separation of C_{12} to C_{20} carboxylic acids on a 10 cm × 1 mm I.D. column packed with Deltabond Octyl (30 nm), **a**, and 10 cm × 1 mm I.D. Nucleosil Cyanopropyl bonded phase (10 nm), **b**, with and without water as a modifier added to supercritical fluid carbon dioxide. Separations were performed at 70 °C with a 2 min hold at 1800 p.s.i. followed by a program of 100 p.s.i./min to 6000 p.s.i. (from Ref. [31])

solvents such as alcohols, amines, and acetonitrile, which do not interfere in the detection of the sample [23, 25, 34, 35].

Water is only sparingly soluble in supercritical fluid carbon dioxide and using a saturator column between the pump and injector, concentrations less than about 0.3% (V/V) in the region of room temperature and normal density operating range can be obtained [31, 32]. The water accumulates in the pores of the packing establishing a steady state relationship with the mobile phase composition. Changes in the equilibrium state relationship with changes in pressure and temperature are fairly rapid allowing the experimental conditions to be quickly optimized in contrast to similar experiences with normal phase HPLC. Figure 4 shows a practical example of the use of water modifier added to carbon dioxide to improve the separation of free fatty acids on two types of column packings, a polymer encapsulated packing (A) and a chemically bonded packing with a low degree of deactivation (B) [31]. In both cases the water modifier is reasonably effective in reducing the silanophilic interactions. Formic acid is a very effective modifier but is limited in general use by its instability and difficulty of purification. Commercially available samples of formic acid have a rather poorly defined composition. Methanol is an effective modifier for masking silanophilic interactions with polar molecules [23, 25]. In this case deactivation may also involve chemical reaction with the silanol groups as discussed previously. For basic samples, alkylamines by themselves, or with other modifiers, are effective masking agents as illustrated by the separation of opium alkaloids in Fig. 5 [32, 35]. Very rapid separations can be obtained on the chemically bonded aminopropylsilanized silica packing, while even bare silica can be dynamically modified to separate the same sample without excessive peak tailing.

Saunders and Taylor [36] have noted a small improvement in the peak shapes of neutral molecules on chemically bonded phases by conditioning

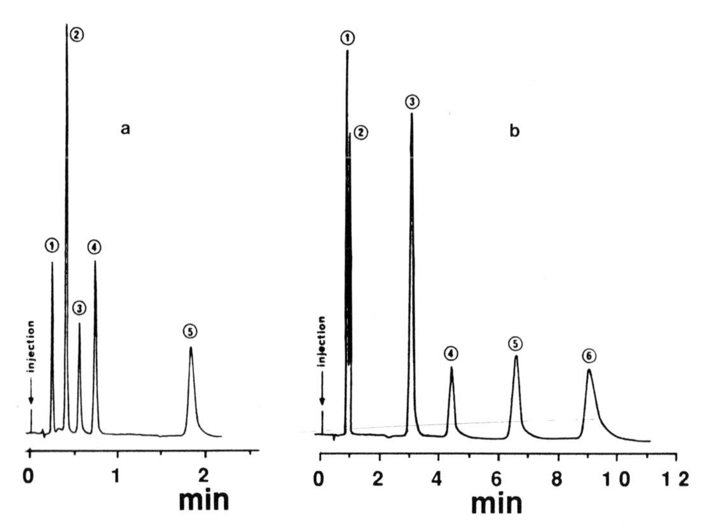

Fig. 5 a, b. Separation of opium alkaloids on a aminopropylsilanized silica packing (Sherisorb-NH_2), 3 μm particle diameter, with carbon dioxide-methanol-triethylamine-water (87.62 : 11 : 80 : 36.0 : 0.22 W/W) as mobile phase, flow rate 8 ml/min, mean pressure 220 bar, temperature = 40.7 °C, detection UV at 280 nm, and column dimensions 12 cm × 0.4 cm I.D. Similar mixture is separated in **b** using naked silica, LiChrosorb Si 60,5 μm particle diameter, and carbon dioxide-methanol-methylamine-water (83.37 : 16.25 : 0.15 : 0.23 W/W) as mobile phase. Other conditions are the same as in **a**. Substance identification for **a**, *1* = narcotine; *2* = thebaine; *3* = codeine; *4* = cryptopine; *5* = morphine: and for **b**, *1* = narcotine; *2* = papaverine; *3* = thebaine; *4* = codeine; *5* = cryptopine; and *6* = morphine (from Ref. [35])

the column at 150 °C prior to use. This they attribute to the removal of adsorbed water. Whether different results might have been obtained for polar molecules is not stated. Their studies do indicate that chemically bonded phases may be used at least up to temperatures of 150 °C without degradation, except for 3-aminopropylsilanized silica packings which showed signs of degradation even at 75 °C.

7.4.3 Shielding of Silanol Groups by Polymer Encapsulation

Exhaustive silanization with the most active of silanizing reagents cannot completely eliminate all silanol groups on the silica surface. Encapsulating the deactivated surface by immobilizing a thin film of a polymeric skin over the surface shields the unbonded silanol groups from the sample and provides reasonably inert column packings [32, 37]. Packings of moderate surface areas based on silicas with 30 nm and 50 nm pore sizes recently became commercially available, as well as a wide range of immobilized ligands including methyl, octyl, octadecyl, phenyl, and cyano with a poly(siloxane) backbone and a poly(ethylene glycol) phase [38]. Although encapsulation shields silanol groups from the sample it is unable to prevent all interactions with the sample because the immobilized film may not cover the whole surface leaving some of the substrate groups exposed and/or because samples soluble in the film must have access to the substrate by diffusion. Since immobilization does not necessarily require chemical bonding to the substrate surface, the general technique can be applied to other substrates beside silica. Alumina-based packings, covered with an immobilized layer of poly-(butadiene), are commercially available. These packings provide good separations of nonpolar compounds, particularly hydrocarbons, but also show undesirable interactions when polar samples are separated. Several research groups are active in the development of immobilized phases on silica and other substrates, and new chromatographic packings can reasonably be expected to emerge from this effort, which will provide a greater shielding of the substrate properties.

7.5 Macroporous Polymeric Packings

Macroporous polymeric packings of different nominal pore sizes based on poly(styrene)-divinylbenzene copolymers have been used as substitutes for silica-based chemically bonded phases when the contribution of silanol groups to retention was found to be undesirable [2, 32, 39–41]. The retention mechanism on these packings seems to involve partitioning with the polymer backbone as well as selective interactions with solutes that contain π electron-acceptor groups [39]. When carbon dioxide is used as the mobile phase retention seems to be high compared to chemically bonded phases, and the efficiency of well retained solutes usually deteriorates compared to earlier eluting peaks. This has tended to limit macroporous polymeric packings to the separation of low-molecular-weight samples. The elution of higher molecular weight samples in an acceptable separation time has been demonstrated using mobile phases containing fairly large amounts of organic solvent modifiers [41]. It is not clear whether high molecular weight samples, say with molecular weights over 1000, can be separated on these columns. There is also little information available concerning the life expectancy of these packings. It has been our experience that during repetitive density programming the constant change in particle dimensions, due to uptake of the mobile phase under supercritical conditions, causes mechanical degradation of the packing.

7.6 Column Packings Used for Special Applications

The separation of racemic mixtures by SFC on chiral phases, originally developed for HPLC, has generated considerable interest in the last few

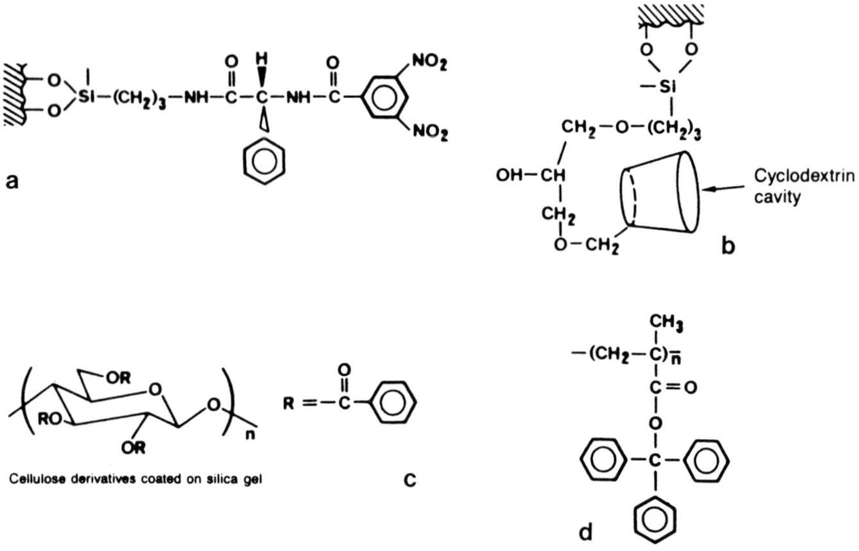

Fig. 6a–d. Structures of some common chiral stationary phases used in SFC. Identification: (**a**) Pirkle phase based on 3,5-dinitrobenzylphenylglycine covalently bonded to 3-aminopropylsilanized silica; (**b**) Cyclobond phase containing cyclodextrin groups chemically bonded to a silica surface; (**c**) Chiralcel OB prepared from the tribenzoate ester of cellulose coated on silica gel; and (**d**) Chiralpak OT (+) consisting of a low molecular weight poly(triphenylmethyl methacrylate) coated on silica gel

years [42–46]. Because of the low viscosity of supercritical fluid mobile phases compared to liquids faster separations are possible under SFC conditions, and this has been the main catalyst for the studies in this area. The stationary phases evaluated include Pirkle phases, cyclodextrins, cellulose esters, helical polymers, and diamide moieties. Some typical examples are shown in Fig. 6. The ionic version of the Pirkle phase in Fig. 6a was unsuitable for SFC because of leaching of the chiral selector from the column under normal operating conditions. The cellulose and polymer phases, Fig. 6c and d, were found to be stable to typical SFC conditions even though the chiral selector is coated and not chemically bonded to a macroporous silica substrate [44, 45]. Most chiral recognition mechanisms depend on multiple polar interactions leading to diastereomeric complexes too stable for the enantiomers to be eluted by carbon dioxide in the absence of polar modifiers. In general, when comparing separations obtained by SFC and HPLC, the same absolute elution order of enantiomers are obtained, but in favorable cases, the resolution per unit time is 5 to 10 times improved in SFC. Optimum resolution was usually observed at low temperatures, frequently subcritical, but greater separation efficiency was found at higher temperatures. Temperature is an important parameter for optimizing resolution in the minimum separation time. For the Pirkle phase there was no significant influence of the average column pressure on selectivity indicating that carbon dioxide does not play a major role in influencing the selectivity of the complex-forming interactions. Separation mechanisms for the cyclodextrin and cellulose phases in SFC, involving both inclusion complexation and intermolecular interactions, paralleled those in HPLC. Dobashi et al. [46] concluded that the use of supercritical carbon dioxide with polar modifiers on chemically bonded diamide phases provided more rapid enantiomer analysis of amino acid derivatives while maintaining the high enantioselectivity under HPLC conditions. It seems clear that SFC is capable of providing faster separations than HPLC but evidence is lacking that SFC can provide separations that could not be obtained by HPLC. A guide to the enantiomeric compounds separated by SFC is given in Ref. [44].

Most of the stationary phases used in contemporary SFC have been chemically bonded or immobilized phases to achieve non-extractability and mechanical stability of the organic film to supercritical fluid carbon dioxide and more polar mobile phases. In a recent study it was shown that liquid organic salts are excellent support deactivating agents with low extractabili-

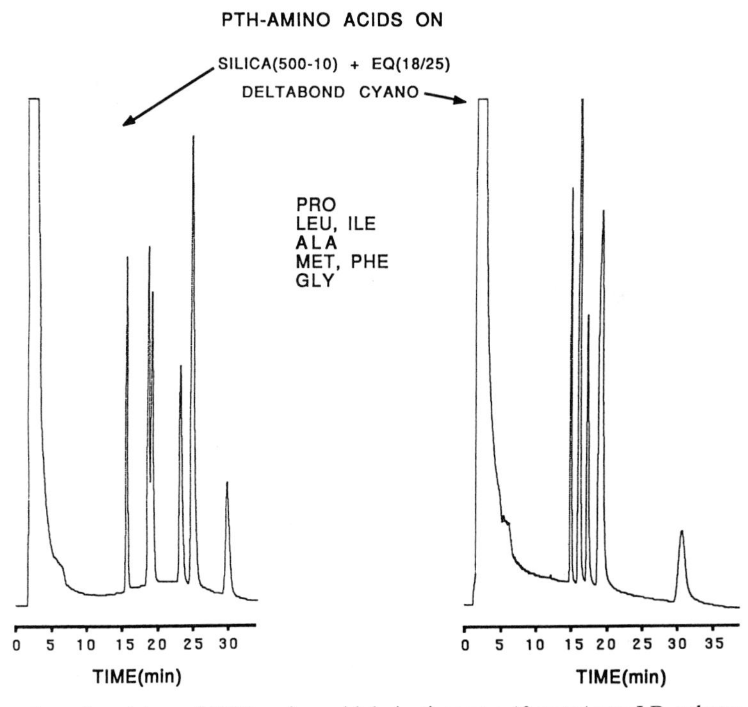

Fig. 7. Separation of a mixture of PTH-amino acid derivatives on a 10 cm × 1 mm I.D. column of Ethoquad (18/25) coated onto Nucleosil 50 nm, 10 µm particle diameter, and a 10 cm × 1 mm I.D. column packed with Deltabond cyanopropyl, 10 µm particle diameter, at 80 °C, with carbon dioxide as mobile phase and a 5 min hold at 80 atm followed by a linear pressure program to 400 atm over 30 min

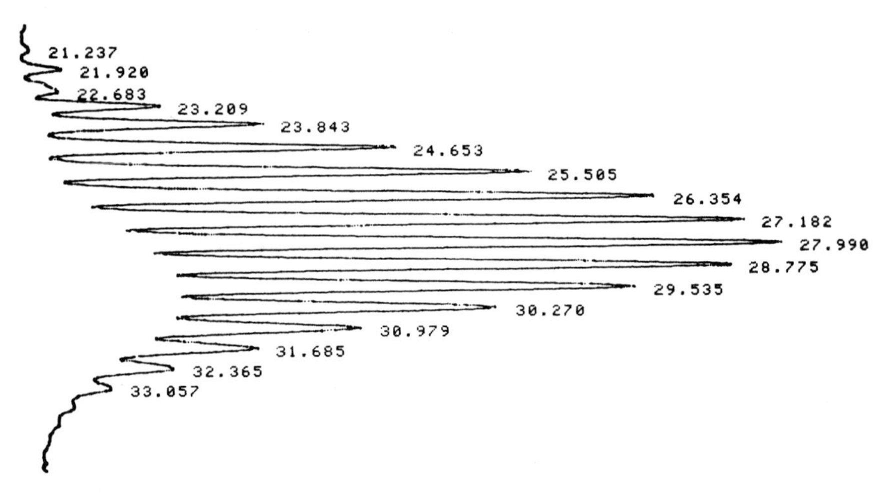

Fig. 8. Separation of Triton X-100 on a 10 cm × 1 mm I.D. column of Nucleosil 50 nm, 10 μm particle diameter, coated with a heavy loading of Ethoquad (18/25) [ca. 20% W/W]. The mobile phase was carbon dioxide, temperature 80 °C, and pressure program, 10 min at 80 atm, increased linearly to 450 atm over 30 min (from Ref. [2])

ty by supercritical fluid carbon dioxide [2]. The liquid organic salts are a new class of polar organic solvents with strong orientation and proton donor/acceptor properties that are easily varied by changing either the anion or cation in a predictable manner [47, 48]. Ethoquad 18/25 (stearylmethyldipolyoxyethylammonium chloride with an average molecular weight of 994) has excellent support deactivating properties in gas chromatography [49] and was used to separate polar solutes by SFC [2]. Figure 7 shows a separation of a mixture of PTH-amino acids on a silica packing coated with Ethoquad 18/25 and the same mixture on a cyanopropyl polymer encapsulated column packing. Both column packings show good peak shape with diminished peak broadening of the PTH-glycine derivative on the Ethoquad (18/25) coated column. Figure 8 shows the separation of Triton X-100 (which has an approximate average molecular weight of 600) on a wide pore silica packing coated with a heavy loading of Ethoquad (18/25). Within the Ethoquad (18/25) coating the Triton X-100 elutes as a very broad peak with virtually no separation of the individual oligomers with supercritical fluid carbon dioxide as the mobile phase. Figure 9 shows the separation of a polarity test mixture on a column of Ethoquad (18/25) when first prepared and again after several months of use. The elution order and peak asymmetry factors for the aged column are not the same as those for the new column, indicating an increase in activity with use. Further evaluation of liquid organic salts, perhaps involving bonding to the silica surface, may lead to improvements in column longevity. They represent a practical alternative to chemically bonded packings and are worthy of further investigation.

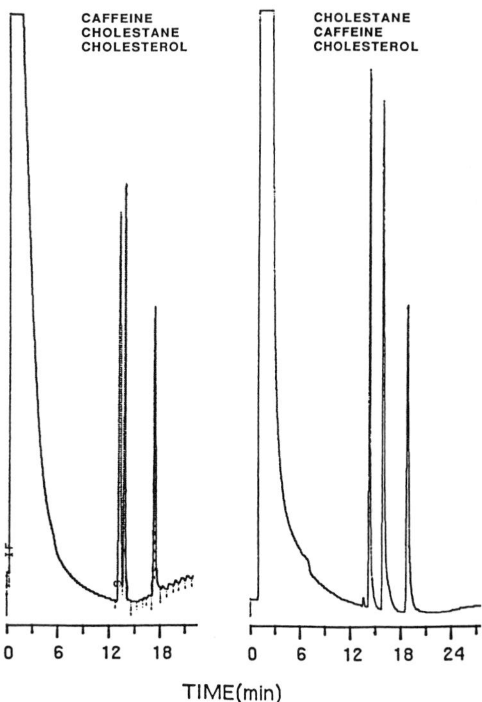

Fig. 9. Separation of a test mixture on a 10 cm × 1 mm I.D. column of 10% (W/W) Ethoquad (18/25) on Nucleosil 50 nm, 10 μm particle diameter, at 100 °C, with carbon dioxide as the mobile phase programmed from 80–480 atm at 20 atm/min

7.7 Conclusions

Stationary phases used in packed column SFC were originally developed for HPLC. This has provided a wide range of selective stationary phases for evaluation and promoted the growth of SFC. The chemically bonded silica based packings provide good separations of nonpolar molecules, but due to the presence of unshielded silanol groups on the packing surface, are not ideal for the separation of many polar samples. Chemically deactivated and polymer encapsulated packings are the most inert packing presently available, but even these will fail in certain critical cases. Further improvements in their chemistry and manufacture are required for the most demanding applications. It is also obvious that the separation of polar samples should be performed on wide pore silica based packings to reduce column activity. The use of polar modifiers in the mobile phase or chemical derivatization to block polar functional groups in the sample are practical methods of circumventing the inadequacy of todays column packings for difficult samples. Derivatization may also increase the solubility of the sample in nonpolar mobile phases, like carbon dioxide, and is worthy of further consideration than it is normally given. A solventless injector for packed column SFC has recently been described that enables volatile derivatizing reagents to be vented from the sample prior to reaching the column, thus avoiding much of the general sample workup, and protecting the column from alteration by the chemically reactive reagents [50]. Macroporous polymeric phases are chemically inert but at present lack mechanical stability and show excessive retention of high molecular weight samples with supercritical carbon dioxide as the mobile phase. The importance of microporosity on their undesirable chromatographic properties in SFC needs to be adequately addressed. The future probably involves the evolution of unique packings for SFC which meet the desired inertness requirements. These may come from the improvement of existing materials, reviewed in this chapter, or just as likely from the specific development of new materials created especially for SFC.

References

1. Poole CF, Schuette SA (1984) Contemporary Practice of Chromatography. Elsevier, Amsterdam
2. Dean TA, Poole CF (1989) J Chromatogr 468:127
3. Sanders LC, Wise SA (1987) CRC Crit Revs Anal Chem 18:299
4. Nawrocki J, Buszewski B (1988) J Chromatogr 449:1
5. Kohler J, Kirkland J (1987) J Chromatogr 385:125
6. Nawrocki J, Moir DL, Szczepaniak W (1989) J Chromatogr 28:143
7. Sands PW, Kim YS, Bass JL (1986) J Chromatogr 360:353
8. Buszewski B, Berek D, Garaj J, Novak I, Suprynowicz (1988) J Chromatogr 446:191
9. Figge H, Deege A, Kohler J, Schomburg G (1986) H Chromatogr 351:393
10. Bien-Vogelsand U, Deege A, Figge H, Koehler J, Schomburg G (1985) Chromatographia 19:170
11. Engelhardt H, Low H, Eberhardt W, Mauss M (1989) Chromatographia 27:535
12. Springston SR, David P, Steger J, Novotny M (1986) Anal Chem 58:997
13. Unger KK (1983) Anal Chem 55:361 A
14. Knox JH, Kaur B, Millward GR (1986) J Chromatogr 352:3
15. Bassler BJ, Hartwick RA (1989) J Chromatogr Sci 27:162
16. Lee DP (1988) J Chromatogr 443:143
17. Stuurman HW, Kohler J, Jansson SO, Litzen A (1987) Chromatographia 23:341
18. Neverjans F, Verzele M (1987) J Chromatogr 406:325
19. King JW, Eissler AL, Friedrich JP (1988) In: Charpentier BA, Sevenants MR (ed) Supercritical Fluid Extraction and Chromatography Techniques and Applications. American Chemical Society, Washington, DC, p 63 (ACS Symposium Series nos 366)
20. Nomura A, Yamada J, Tsunoda K-I (1988) J Chromatogr 448:87
21. Nomura A, Yamada J, Tsunoda K-I, Sakaki K, Yokochi T (1989) Anal Chem 61:2076
22. Warren FV, Bidlingmeyer B (1984) Anal Chem 56:950
23. Schoenmakers PJ, Uunk LG, De Bokx PK (1988) J Chromatogr 459:201
24. Ashraf-Khorassani M, Taylor LT (1988) J Chromatogr Sci 26:33
25. Ashraf-Khorassani M, Taylor LT (1989) J High Resolut Chromatogr 12:40
26. Doehl J, Farbrot A, Greibrokk T, Iversen B (1987) J Chromatogr 392:175
27. Jinno K, Niimi S (1988) J Chromatogr 455:29

28. Evans MB, Smith MS, Oxford JM (1989) J Chromatogr 379:170
29. Hirata Y (1984) J Chromatogr 315:31, 39
30. Schmitz FP, Leyendecker D, Leyendecker D (1987) J Chromatogr 389:245
31. Geiser FO, Yocklovich SG, Lurcott SM, Guthrie JW, Levy EJ (1988) J Chromatogr 459:173
32. Engelhardt H, Gross A, Mertens R, Petersen M (1989) J Chromatogr 477:169
33. Schwartz HE, Barthel PJ, Moring SE, Yates TL, Lauer HH (1988) Fresenius' Z Anal Chem 330:204
34. Blilie AL, Greibrokk T (1985) Anal Chem 57:2239
35. Janicot JL, Caude M, Rosset R (1988) J Chromatogr 437:351
36. Saunders CW, Taylor LT (1989) Chromatographia 28:253
37. Ashraf-Khorassani M, Taylor LT, Henry RA (1988) Anal Chem 60:1529
38. Keystone Scientific Newsletter Vol 2, nos 2 (1989)
39. Morin P, Caude M, Rosset R (1987) J Chromatogr 407:87
40. Smith RM, Sanagi MM (1989) J Chromatogr 481:63
41. Gemmel B, Lorenschat B, Schmitz FP (1989) Chromatographia 27:605
42. Mourier PA, Eliot E, Caude MH, Rosset RH, Tambute AG (1985) Anal Chem 57:2819
43. Hara S, Dobashi A, Hondo T, Saito M, Senda M (1986) J High Resolut Chromatogr 9:249
44. Macaudiere P, Caude M, Rosset R, Tambute A (1989) J Chromatogr Sci 27:383
45. Macaudiere P, Caude M, Rosset R, Tambute A (1989) J Chromatogr Sci 27:583
46. Dobashi A, Dobashi Y, Ono T, Hara S, Saito M, Higashidate S, Yamauchi Y (1989) J Chromatogr 461:121
47. Poole CF, Furton KG, Kersten BR (1986) J Chromatogr Sci 24:400
48. Kersten BR, Poole SK, Poole CF (1989) J Chromatogr 468:235
49. Furton KG, Poole SK, Poole CF (1987) Anal Chim Acta 192:49
50. Dean TA, Poole CF (1989) J High Resolut Chromatogr 12:773

8 Enantiomer Separation by Capillary Supercritical Fluid Chromatography

MICHAEL SCHLEIMER and VOLKER SCHURIG

8.1 Introduction

The unambiguous determination of enantiomeric compositions and absolute configurations is an important analytical task in the synthesis, characterization and use of chiral nonracemic compounds (optical isomers, enantiomers) such as chiral auxiliaries, catalysts, pharmaceuticals, pesticides, herbicides, pheromones, flavours and fragrants. As the insight into chirality-activity relationships steadily improve and, as a consequence, legislation of chiral drugs becomes more and more stringent, the development of precise methods for the determination of enantiomeric purities up to ee > 99% is of great importance.

Enantiomer analysis by chiroptics (polarimetry, circular dichroism) or spectroscopy (nuclear magnetic resonance) is unsuitable for the precise determination of enantiomeric impurities lower than 1% while chromatographic methods can usually cope with this requirement at low sample sizes [1]. In chromatography, the separation of enantiomers relies on the interaction between a chiral (nonracemic) selector and the chiral selectand in the spirit of Pasteurs resolution principles via diastereomers. Three different approaches, suitable for SFC, can be distinguished [2]:

- Preformation of diastereomers by off-column derivatization and use of an achiral stationary phase
- Intermediate formation of diastereomers via chiral mobile phase additives and use of an achiral stationary phase
- Intermediate formation of diastereomers via the use of a chiral stationary phase (CSP).

Off-column derivatization is described in [3–5]. An example is shown in Fig. 1, where the esterification of a racemic secondary alcohol with (*S*)-Trolox™ and the separation of the preformed diastereomers was carried out on an achiral stationary phase. An investigation on the use of chiral mobile phase additives for the enantiomer separation of 1,2-amino-alcohols in packed-column-SFC has also been described [6]. The application of CSP for the direct enantiomer separation of (preferably underivatized) solutes by SFC is straightforward and will be discussed herein.

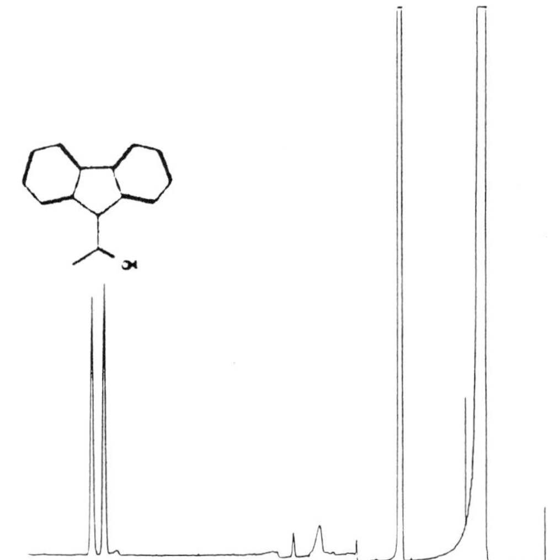

Fig. 1. SFC chromatogram [5] of diastereomeric derivatives of methyl-9-*H*-fluorenyl-9-methanol on a 10 m ×50 μm i.d. fused silica column, coated with a poly (30%-biphenyl)-methylsiloxane (d_f = 0.25 μm). 100 °C, CO_2 with density program from 0.2 g ml^{-1} to 0.75 g ml^{-1} at 0.01 g ml^{-1} min^{-1}; FID

8.2 General Aspects

The rapid improvement of column technology and instrumentation in the past decade as well as the better understanding of the physical chemistry of supercritical fluids render SFC an attractive complementary method for the separation of enantiomers. There are several specific options for the use of SFC.

8.2.1 Temperature

In enantiomer separation by chromatography, the *physical* partition equilibrium between the mobile and stationary phase is superimposed by a fast and reversible *chemical* association equilibrium of the selectand and the selector. Thus, the observed difference in retention of the enantiomers is caused by the different stability of the diastereomeric associates of the racemic selectand and the non-racemic selector, which depends on the concentration and the enantiomeric purity of the selector as well as the temperature. The thermodynamic description of this process according to the Gibbs-Helmholtz relationship

$$RT \ln (K_R/K_S) = -\Delta_{R,S}(\Delta G°) = -\Delta_{R,S}(\Delta H°) + T\Delta_{R,S}(\Delta S°)$$

implies the existence of a temperature T_{iso} (isoenantioselective temperature), where the entropy and enthalpy contributions to chiral recognition cancel each other to the effect that enantiomer separation can not be observed at any circumstance [7–9].

$$T_{iso} = \Delta_{R,S}(\Delta H^\circ)/\Delta_{R,S}(\Delta S^\circ) \quad \text{for} \quad \Delta_{R,S}(\Delta G^\circ) = 0 \ .$$

At T_{iso} the elution order of the enantiomers is reversed and above T_{iso} enantiomer discrimination is entropy-controlled and separation factors α will increase with increasing temperature while below T_{iso} enantiomer discrimination is enthalpy-controlled and separation factors α will increase with decreasing temperature [10]. Besides some rare cases in GC [9] and LC (e.g. ligand exchange chromatography) [11] enantiomer separation is usually governed by enthalpy-control. Consequently, it must be the aim to lower the temperature of analysis in order to enhance enantioselectivity.

Another, more practical reason for using lower temperatures may arise from the (limited) thermal stability of either the CSP or the solute to be separated. Column bleeding at high temperatures, due to degradation of the CSP, as well as racemization at the chiral centers of the selectand or selector may be another reason to avoid, e.g., temperature programming as employed in HRGC. The use of SFC may also circumvent the need of solute derivatization with its inherent disadvantages [12, 13] which is often required in GC to improve volatility, enantioselectivity or detectability. In SFC, enantiomers can be eluted at moderate temperatures without the need of derivatization. The loss of efficiency as the result of lower diffusion coefficients at low temperatures in SFC is often compensated by a strong enhancement of the enantioselectivity as expressed by the separation factor α.

8.2.2 Speed of Analysis and Efficiency

Due to lower volumetric flow rates and higher diffusion coefficients of solutes in supercritical fluids, the speed of analysis is increased in SFC compared to LC without affecting efficiency. This may be one of the greatest advantages in the use of packed-column SFC, i.e., resolution per unit time is often increased significantly. An example [14] can be seen in Fig. 2, where resolution was set constant but runtime decreased by a factor of 4–5 under subcritical conditions, compared to LC. Although SFC cannot compete with GC in regard to speed of analysis and efficiency, the use of small-diameter WCOT-columns producing high plate numbers is common to SFC, but not to LC. Due to the greater permeability of fluids in the supercritical state column length can be increased and, consequently, racemates with small separation factors α can be resolved. Besides the low number of active sites at the inner surface of the fused-silica column, capillary SFC has the additional advantage of requiring low flow rates compared to packed-column techniques. This permits not only the application of flow-sensitive detectors such as the linear, universal FID or MS-coupling techniques producing high sensitivities from narrow peaks, but also the extensive use of

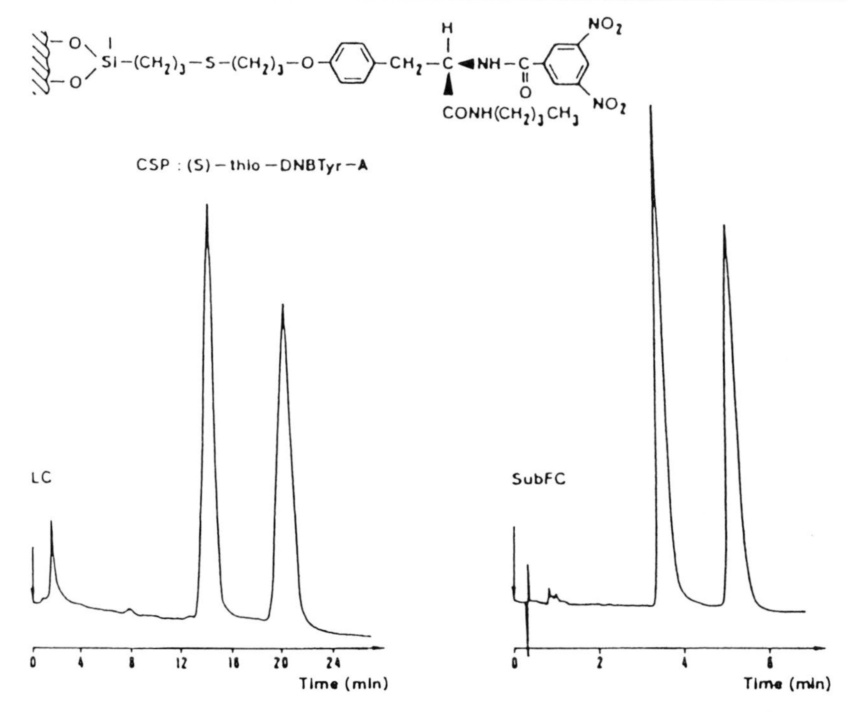

Fig. 2. Comparison of the analysis times in LC and sub-SFC [14] for oxazepam enantiomers. Column: 15 cm × 4.6 mm i.d., 5 μm. LC: *n*-hexane/ethanol 90 : 10 (v/v); flow rate 2 mL min^{-1}; 25 °C; SFC: CO_2/ethanol 92 : 8 (w/w); flow rate 6 mL min^{-1} at 0 °C; average column pressure 20 MPa; 25 °C; UV detection at 229 nm

density programming. The efficiency of an enantiomer separation is not only a function of the chromatographic variables, but is also strongly influenced by the temperature-dependent kinetics of the formation of the diastereomeric associates, which must be fast and reversible. If the kinetics of the interaction between the selectand and selector are slow in terms of the chromatographic time scale involving one theoretical plate, peak shapes of the enantiomers will be distorted or may even show great differences in peak width, leading to a reduced efficiency (cf. Fig. 3).

The quality of a successful enantiomer separation in SFC depends on several variables and is still in an empirical state. In practice, the separation factor α, reflecting the difference of the free energy of diastereomeric association, should not be used as a single criterion for the quality of enantiomer separation, because efficiency is not considered and it is not readily applicable in temperature or pressure programmed runs. A more practicable criterion for a quantitative enantiomer separation is the resolution

$$R_s = \frac{\sqrt{N}}{4}\left(\frac{\alpha-1}{\alpha}\right)\left(\frac{k_2'}{k_2'+1}\right), \tag{1}$$

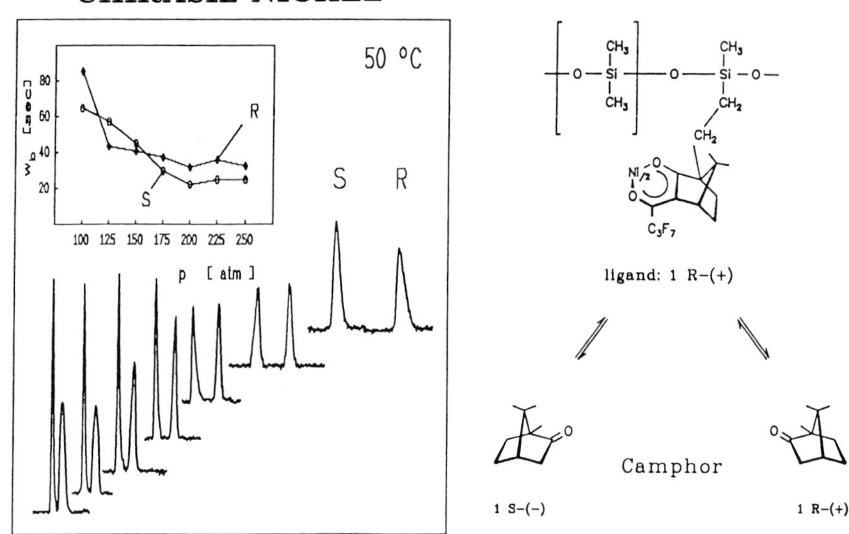

Fig. 3. Variation of peak width on applied pressure for racemic camphor, measured at 50 °C on a 2 m×50 μm i.d. fused silica column, coated with Ni(II)-Chirasil-Metal ($d_f = 0.25$ μm). From 10 MPa (*right*) to 25 MPa (*left*), measured in 2.5 MPa steps, peak width of both enantiomers alternate

or the separation number (Trennzahl) introduced by Kaiser [15]. A similar factor cR (chiral resolution) was recently defined by Aichholz et al. [16]. These values also consider the efficiency of the system and can be used for (enriched) racemates even in programmed runs.

8.2.3 Mobile Phase

Chromatographically important properties of supercritical fluids are the density, viscosity and the diffusion coefficient of solutes [17]. The elution of enantiomers in SFC is commonly maintained by increasing the density of the mobile phase during an isothermal run, which simultaneously increases the solubility of the solutes. This is easily achieved by varying the applied pressure, thus permitting the elution even of enantiomers with high molecular weights and low volatilities. As carbon dioxide, which is most commonly used as sub- or supercritical mobile phase, has no permanent dipole, but is polarizable, changes in the solvation strength during density programming are rather small. The addition of polar modifiers, applied often in packed column-SFC, enhances not only the solvation strength of the mobile phase, but may also diminish the mixed retention-mechanism especially of polar solutes by blocking active sites at the surface of the packing material or the column wall in capillaries. The existence of residual silanol groups on coated silica packings is one of the greatest disadvantages of packed-column SFC, analogous to LC [18, 19]. The blocking of these active sites by polar modifiers influences not only enantioselectivity (because modifiers compete with the solute for the chiral sites of the stationary phase), but it affects also the critical parameters of the mobile phase, frequently forcing the performance of chromatography to the less efficient subcritical region.

The increase of the mobile phase density and/or polarity increases both the solute interactions with the mobile phase [20, 21] and the interaction of the supercritical fluid with the chiral stationary phase, thus influencing the difference of the free enthalpy of the diastereomeric associates either by enthalpic or entropic effects. In extreme cases, the mechanism of chiral recognition may be altered to such an extent that, as it was shown in LC, the change of the mobile phase led to a reversed elution order of the enantiomers [22].

The complexity of the influence of the mobile phase density and composition upon the chiral discrimination process complicates the both theoretically and practically interesting comparison of the different methods (LC, SFC, GC). From a more theoretical point of view, enantiomer separations performed by SFC can be considered as a compromise between the high efficiencies of GC and the solubilizing properties of liquid mobile phases of HPLC, which, in combination with low temperatures, should result in high enantioselectivities and efficiencies lying between LC and GC. In practice, the performance strongly depends on the CSP applied and, among others, upon its compatibility with the mobile phase for many classes of compounds. For instance, cyclodextrin selectors displaying enantioselectivity for numerous racemates [23], require operation in the reversed phase mode in most chiral LC applications since the hydrophobic cavity of the cyclodextrin interacts strongly with nonpolar mobile phases. Because of its moderate polarity, supercritical carbon dioxide may be able to substitute most of the normal phase LC separations.

Compared to conventional LC it is generally observed in packed-column SFC that enantioselectivity is lower or similar, but efficiency as well as speed of analysis is increased. Compared to GC, capillary SFC usually shows a more dramatic loss in enantioselectivity and efficiency, but the great differences in the analysis temperature available to both methods often allows SFC to overcome these disadvantages.

8.2.4 Column Loadability

Generally, because of the chemical nature of the recognition process, chiral separation systems are more sensitive to solute overloading phenomena than achiral systems, whereby the capacity depends strongly on the total amount of the chiral selector in the column.

In packed columns, in which the chiral selector is merely "diluted" by its own spacer, attaching it to the packing material, problems with solute overloading are less pronounced. Consequently, semi-preparative scale enantiomer separations are feasible [24, 25].

In capillary SFC all CSP known today are based on polysiloxanes. Thus, the chiral selector is "diluted" to about 5% – 40% by weight of the total stationary phase. Column loadability decreases also drastically with the use of small-diameter columns, which are more strongly recommended in SFC than in GC due to their higher efficiency. Loadability is normally increased by using thick films (up to 1 μm) of stationary phase [26].

8.3 Packed-Column SFC

Since 1985, when the first enantiomer separation in packed-column SFC using a Pirkle-type CSP was reported by Mourier et al. [27], the use of commercially available LC-columns has become quite popular in SFC. This led to numerous applications on a great variety of CSPs, which were reviewed by Macaudiere et al. [28] in 1989.

Later, Dobashi et al. [29] prepared a novel valine-diamide phase for the separation of derivatized amino acids. The influence of end-capping of the remaining surface silanols on resolution and amount of modifier was discussed also.

Recently, Nitta et al. [30] reported on the use of cellulose-tris(phenylcarbamate) for the separation of stilbene oxide enantiomers in LC and SFC, whereas Gasparrini et al. [31] investigated a 3,5-dinitrobenzoyl-(R, R)-1,2-diaminocyclohexane derivative, bonded to glycidoxypropylated silica, for enantiomer separation of racemic sulfoxides, flurbiprofen, and alprenolol derivatives. Separations were achieved using a column of 150 × 1.2 mm i.d., packed with 5 μm particles.

8.4 Open Tubular Column SFC

The first separation of chiral compounds in capillary SFC was reported by Röder et al. [32] in 1987. They also applied a Pirkle-type selector, which was anchored to an immobilizable polysiloxane, for the separation of several derivatized amino acids (cf. Fig. 4). The comparison of selectivities under SFC (capillary) and LC (packed) conditions, which – as pointed out before – is not unambiguous, showed smaller separation factors under supercritical conditions. This was explained by the higher temperatures used in SFC as well as by the higher polarity of CO_2 (which may itself undergo in-

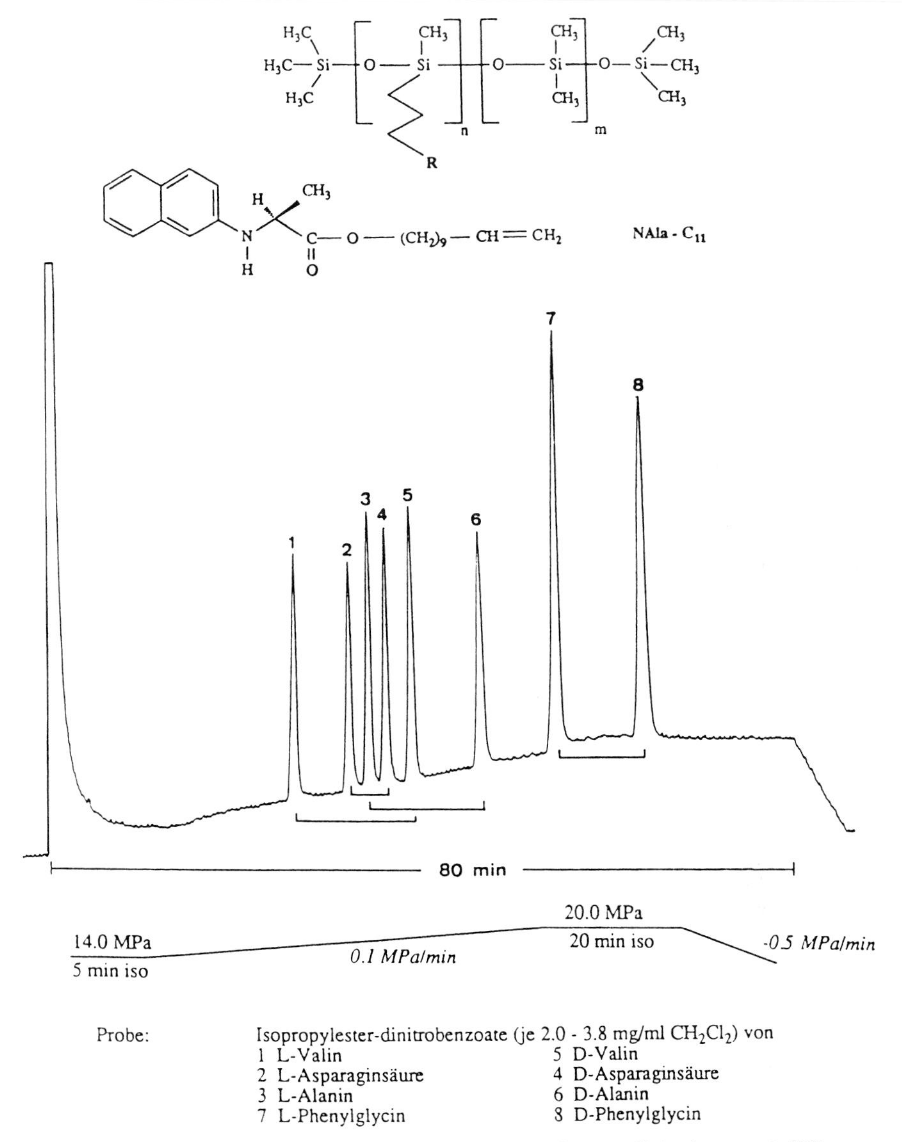

Fig. 4. Enantiomer separation of amino acids (isopropyl ester, dinitrobenzoate) [32] on a 5 m × 50 μm i.d. capillary column, coated with a Pirkle-type derivatized polymethylsiloxane ($d_f = 0.25$ μm). 80 °C, CO_2 with pressure program from 14.0 MPa (5 min hold) to 20.0 MPa at 0.1 MPa min^{-1}; FID

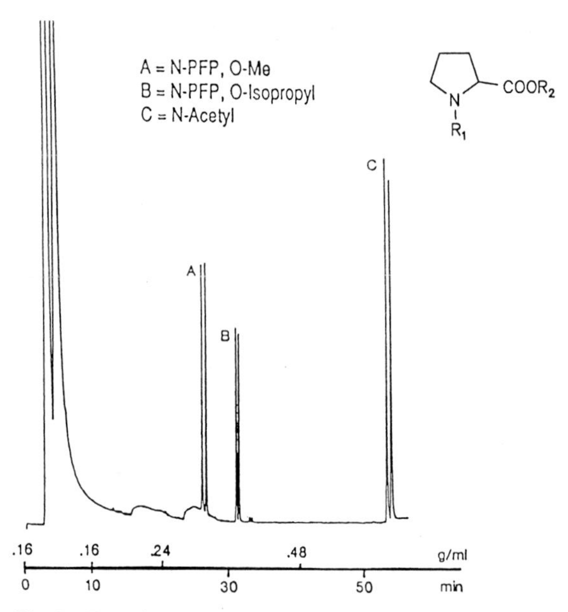

Fig. 5. Enantiomer separation of different derivatives of racemic proline [34] on a 10 m × 50 μm i.d. capillary column, coated with Chiral-NEB (d_f = 0.15 μm). 60 °C, CO_2 with density program from 0.16 g ml^{-1} (10 min hold) to 0.48 g ml^{-1} at 0.006 g ml^{-1} min^{-1}; FID

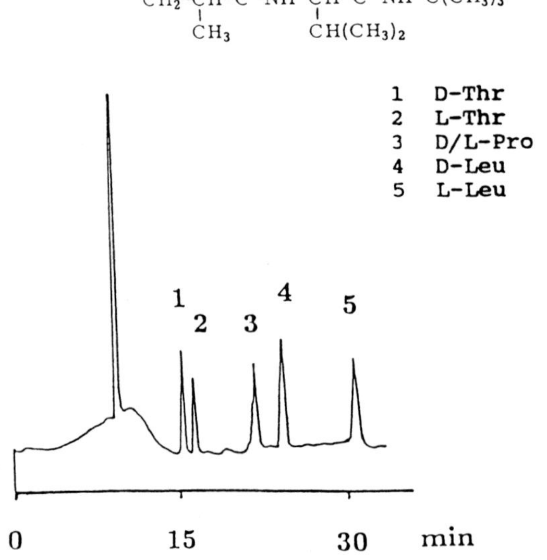

Fig. 6. Enantiomer separation of amino acids (*n*-propyl ester, N-TFA) [36] on a 10 m × 100 μm i.d. capillary column, coated with Chirasil-Val ($d_f \approx$ 0.25 μm) at 50 °C. CO_2 at 7.7 MPa; FID

teractions with the chiral selector) as compared to *n*-heptane and added modifier used in normal phase HPLC.

Bradshaw et al. [33] synthesized polysiloxanes containing pendant chiral amide side chains and Rouse [34] used that derived from (*S*)-1-(1-naphthyl)-ethylamine for the separation of racemic proline with different derivatization in capillary SFC (cf. Fig. 5).

Chirasil-Val, a well-known CSP for the enantiomer separation of derivatized amino acids in GC [35], was used in SFC by Lai et al. [36]. Significantly lower enantioselectivities were found in SFC compared to GC at the same temperature. A representative chromatogram obtained on a 100 μm i.d. column is shown in Fig. 6. The loss in enantioselectivity was ascribed to an inhibitory effect of solvated solute and selector during the diastereomeric interaction by strong hydrogen bonding with the supercritical mobile phase carbon dioxide. Another group [37, 38] obtained contrary results in the use of OV225-L-valine-tert-butylamide as CSP under GC and SFC conditions.

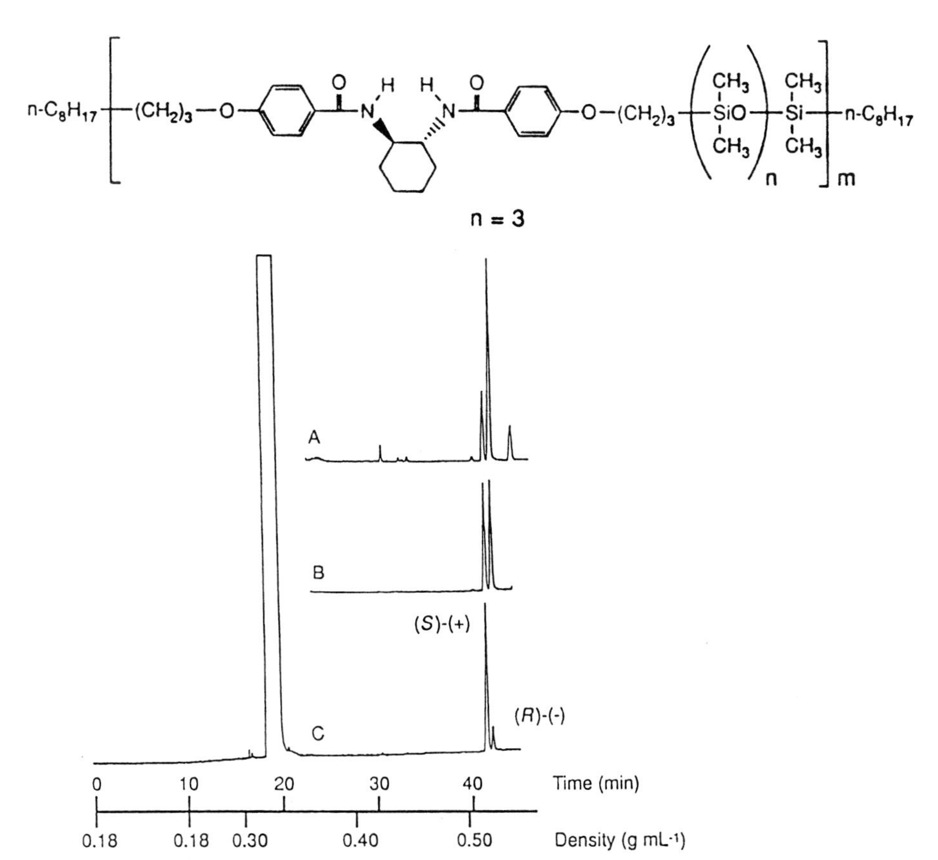

Fig. 7. Enantiomer separation of 3,3-dimethyl-butane-1,2-diol [39] on a 20 m × 50 μm i.d. capillary column, coated with a poly-(cyclohexyldiamide)-methylsiloxane ($d_f = 0.20$ μm). 50 °C, CO_2 with density program from 0.18 g ml^{-1} (10 min hold) to 0.30 g ml^{-1} at 0.02 g ml^{-1} min^{-1} and to 0.90 g ml^{-1} at 0.009 g ml^{-1} min^{-1}; FID

Petersson et al. [39] and Rossiter et al. [40] developed a new strategy for the synthesis of chiral polysiloxanes. They used block polymers, in which the polysiloxane units are connected via a chiral cyclohexylene-bis-benzamide derivative. Several diols were separated (cf. Fig. 7) with high efficiency on a long (20 m × 50 μm) capillary column.

Brügger et al. [41] and Marti et al. [42] more recently described the use of CSPs derived from (*R*)-*N*-pivaloyl-naphthylethylamide or (*R*)-*N*-(1-phenylethyl)-*N'*-[3-(diethoxymethylsilyl)-propyl]urea for enantiomer separation of derivatized amines in SFC and GC.

Schurig et al. used an immobilized (non-extractable) CSP. Consisting of a permethylated monokis-(6-O-octamethylene)-β-cyclodextrin chemically bonded to a dimethylpolysiloxane (Chirasil-Dex) in both GC [43] and SFC [44]. The enantiomer separations of underivatized pharmaceuticals such as syncumar, dihydrodiazepam and ibuprofen (a free acid) were achieved on a short (2.5 m × 50 μm i.d.) nondeactivated fused silica capillary column (cf. Fig. 8). Although enantioselectivity as expressed by the separation factor α is lower than in most of the LC separations, the high resolution obtained in combination with the use of the universal FID bears certain advantages in the trace analysis of enantiomeric excess greater than 99%. A systematic comparison of enantiomer separation in inclusion-GC and SFC [45] showed a dramatic loss of enantioselectivity under supercritical conditions at a given temperature. This was attributed to the blocking of the hydrophobic cyclo-

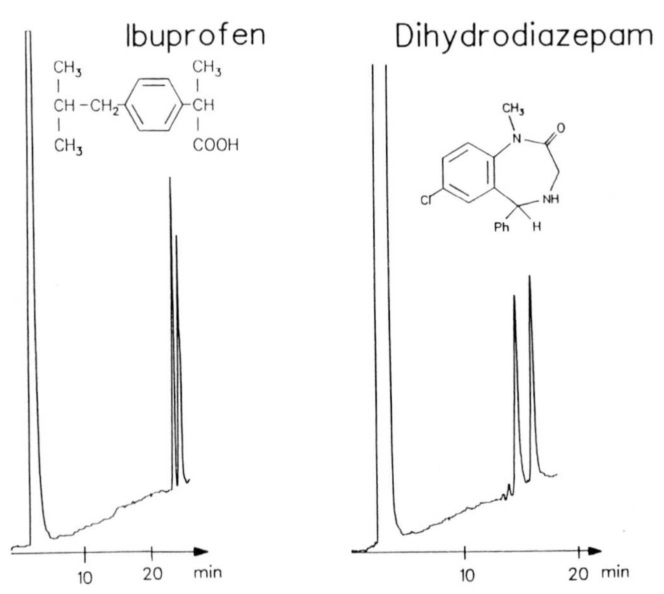

Fig. 8. Enantiomer separations of pharmaceuticals [44] on a 2.5 m × 50 μm i.d. Chirasil-Dex coated fused silica column ($d_f = 0.25$ μm). *Left:* Ibuprofen; 60 °C, CO_2 with density program from 0.25 g ml^{-1} (2 min hold) at 0.0035 g ml^{-1} min^{-}. *Right:* Dihydrodiazepam; 90 °C, CO_2 with density program from 0.31 g ml^{-1} (2 min hold) at 0.0029 g ml^{-1} min^{-1}

dextrin cavity by nonpolar carbon dioxide molecules at high densities. The effect of temperature on resolution was also studied (cf. Fig. 9), showing the same typical maximum which is observed for the capacity factors of the enantiomers. Generally, a better resolution is obtained at a lower inlet pressure (density).

More recently, Schurig et al. [46] reported on the use of a polysiloxane based CSP, containing a chiral metal chelate (Chirasil-Metal) in capillary GC and SFC. In complexation chromatography enantiomer separation is achieved by donor-acceptor interactions between a chiral solute possessing

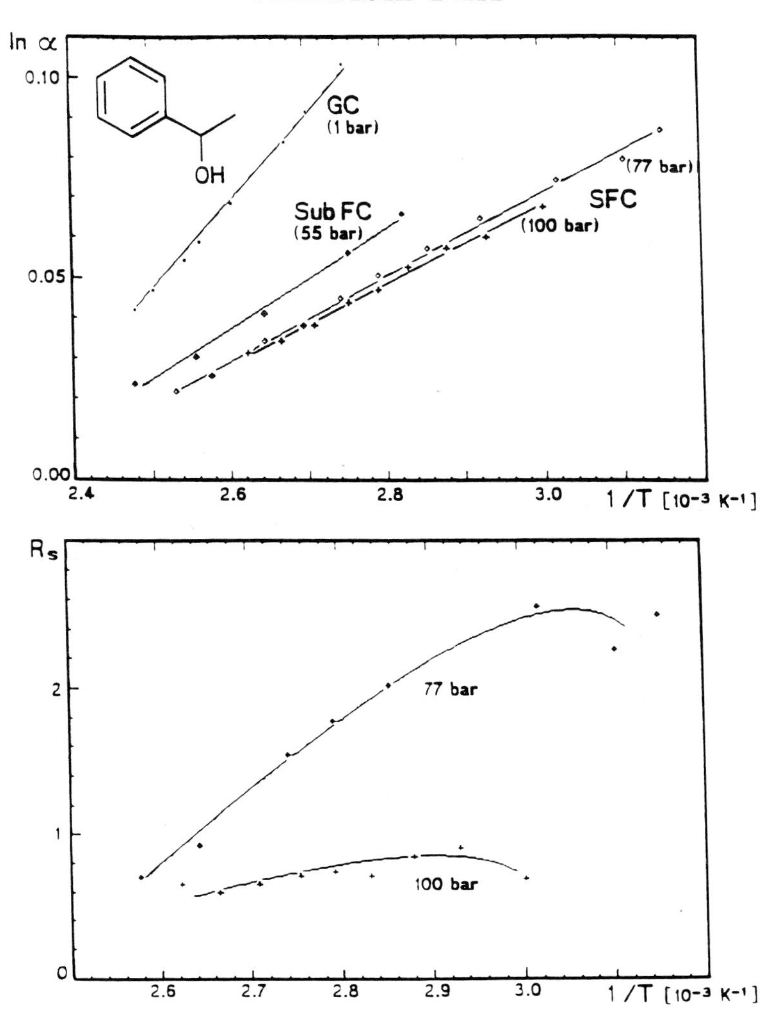

Fig. 9. GC-SFC-comparison of separation factors α and resolution R_s as function of temperature for 1-phenylethanol in *inclusion chromatography* [45] using a 10 m×100 μm i.d. fused silica column, coated with Chirasil-Dex (d_f = 0.25 μm)

lone electron pairs and a sterically and electronically unsaturated metal chelate bearing non-racemic terpeneketonate ligands such as nickel(II)-bis[(3-heptafluorobutanoyl)-(1*R*)-camphorate]. Thus enantioselectivity of this system is strongly affected by variations of either the ligand (e.g. derivatives of camphor, carvone, pulegone or menthone) or the Lewis acidity of the metal ion [M(II), e.g., Ni(II), Mn(II), Cu(II), Zn(II)]. The standard system, recently applied, is a polysiloxane anchored Ni(II)-bis[(3-heptafluorobutanoyl)-(1*R*)-camphorate], Chirasil-Nickel. Chromatograms obtained by complexation SFC on Ni(II)-, Mn(II),- and Zn(II)-based Chirasil-Metals [47] are shown in Fig. 10.

The Chirasil-Nickel-based CSP shows no loss in enantioselectivity over the whole range of temperature and density applied (cf. Fig. 11), thus indicating that blocking of the chiral selector by supercritical carbon dioxide plays no negative role. In some cases higher separation factors are observed in SFC as compared to GC. At constant resolution (e.g. $R_s = 2.0$, measured for 1-phenylethanol), the analysis temperature decreases with increasing pressure from 145 °C in GC (1 atm N_2) to 45 °C in SFC (250 atm CO_2). As pointed out before, the dramatic loss of efficiency in SFC as compared to GC can thus be compensated by applying low-temperature SFC.

8.5 Conclusion

The separation of enantiomers on chiral stationary phases (CSP) by SFC has been demonstrated by *Π-Π*-donor-acceptor interaction, hydrogen-bonding, inclusion and complexation. It is believed that SFC will find an established place in modern enantiomer analysis as instrumentation and the availability of hard- and software improve further. Although SFC is unlikely to substitute either GC or LC in enantiomer analysis, it represents a versatile complementary tool combining advantages of both classical methods.

Note added in proof: A comprehensive review on enantiomer separation by SFC has recently been published by Juvancz and Markides in LC-GC (Vol. 5 (1992) 44).

Fig. 10. Enantiomer separations of different solutes by *complexation SFC* [47]. *Upper:* Enantiomer separations of underivatized secondary alcohols on a 2 m×50 μm i.d. fused silica column, coated with Chirasil-Nickel ($d_f = 0.25$ μm). 120 °C, CO_2 with density program as indicated. *Middle:* Enantiomer separation of 2,2′-(dihydroxymethylene)-1,1′-binaphthyl on a 4 m×50 μm i.d. fused silica column, coated with Chirasil-Manganese ($d_f = 0.25$ μm). 50 °C, CO_2 with density program as indicated. *Lower:* Enantiomer separation of racemic benzoine on a 2.5 m×50 μm i.d. fused silica column, coated with Chirasil-Zinc ($d_f = 0.25$ μm). 70 °C, CO_2 with density program as indicated; FID

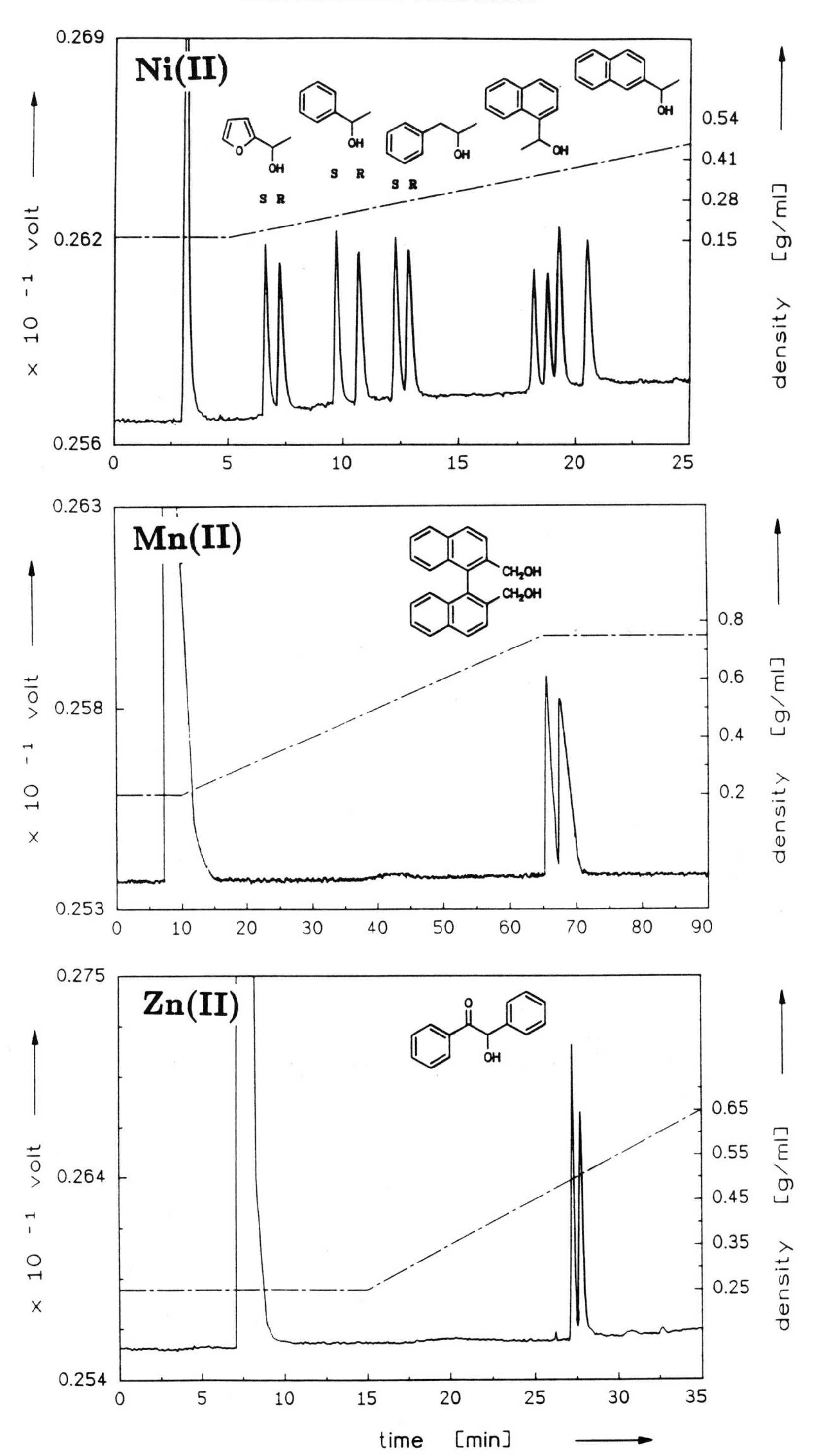
CHIRASIL-METAL
Ni(II)
Mn(II)
Zn(II)
x 10 -1 volt
density [g/ml]
time [min]
S R
S R
S R
OH
CH2OH
CH2OH
0.269
0.262
0.256
0.54
0.41
0.28
0.15
0.263
0.258
0.253
0.8
0.6
0.4
0.2
0.275
0.264
0.254
0.65
0.55
0.45
0.35
0.25

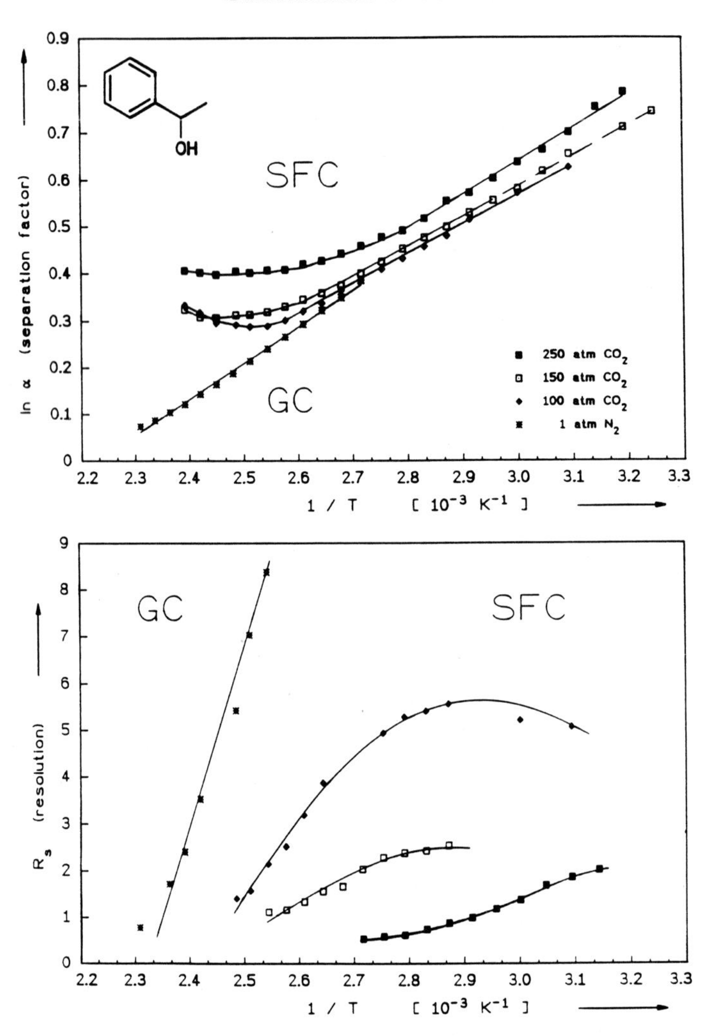

Fig. 11. GC-SFC Comparison of separation factors α and resolution R_s as function of temperature for 1-phenylethanol in *complexation chromatography* [47], using a 2 m×50 μm i.d. fused silica column, coated with Chirasil-Nickel (d_f = 0.25 μm)

References

1. Schurig V (1985) Kontakte (Darmstadt) 1 : 54 (Part 1) and (1985) 2 : 22 (Part 2) and (1986) 1 : 1 (Part 3)
2. a) Pirkle WH (1983) In Asymmetric Synthesis; Morrison JD (ed) (1983) Academic Press, Inc.: New York; Vol 1 Chapter 6; b) Schurig V in Asymmetric Synthesis; Morrison JD (ed) (1983) Academic Press, Inc: New York, Vol 1, Chapter 5; c) Allenmark SG (1988) in Chromatographic Enantioseparation; John Wiley, New York; d) Krstulovic AM (1989) in Chiral Separations by HPLC. John Wiley, New York
3. Sakkai K, Hirata H (1991) J Chromatogr 585:117
4. Walther W, Vetter V, Netscher T (1992) J Microcol Sep 4:45
5. Analytical Supercritical Fluid Chromatography and Extraction; Lee ML and Markides KE (eds) (1990) Chromatography Conferences, Inc: Provo, Utah; p 544
6. Steuer W, Schindler M, Schill G, Erni F (1988) J Chromatogr 447:287
7. Koppenhoefer B, Bayer E (1984) Chromatographia 19:123
8. Schurig V (1984) Angew Chem Int Ed Engl 23:747
9. Schurig V, Ossig A, Link R (1989) Angew Chem 101:197
10. Reversal of the elution order of the enantiomers may also be due to mechanistical changes, as it was found e.g. in Ref [14]

11a. Darahkov VA, Kurganov AA (1987) Adv Chromatogr 22:17

11b. Gübitz G (1986) J Liq Chromatogr 90:516

12. Knapp DR (1979) in Handbook of Analytical Derivatization Reactions. John Wiley and Sons, New York
13. McKenzie SL, Tenaschuk D (1979) J Chromatogr 173:53
14. Macaudiere P, Caude M, Rosset R, Tambute A (1989) J Chromatogr Sci 27:583
15. Kaiser RE (1976) Chromatographia 9:337 and (1977) 10:445
16. Aichholz R, Bölz U, Fischer P (1990) J High Resolut Chromatogr 13:234
17. Bartle KD (1988) in Supercritical Fluid Chromatography (Smith RM, ed). The Royal Society of Chemistry, London
18. Nawrocki J (1991) Chromatographia 31:177
19. Pirkle WH, Readnour RS (1991) Chromatographia 31:129
20. Mulcahey LJ, Taylor LT (1990) J High Resolut Chromatogr 13:393
21. Knowles DE, Richter BE (1991) J High Resolut Chromatogr 14:689
22. Macaudiere P, Lienne M, Caude M, Rosset R, Tambute A (1989) J Chromatogr 467:357
23. Schurig V, Novotny H-P (1990) Angew Chem 102:969
24. Pericles H, Giorgetti A, Dätwyler P in: Markides KE, Lee HL (eds) SFC Applications Handbook. Brigham Young University Press, Utah, Provo 1988, p 44
25. Fuchs G, Doguet L, Perrut M, Tambute A, LeGoff P (1991) in Proceedings of 2nd International Symposium on SFC, Boston, May 19–20
26. Peaden PA, Lee ML (1983) J Chromatogr 259:1
27. Mourier PA, Eliot E, Caude MH, Rosset RH, Tambute AG (1985) Anal Chem 57:2819
28. Macaudiere P, Caude M, Rosset R, Tambute A (1989) J Chromatogr Sci 27:383
29. Dobashi A, Ono T, Hara S, Saito M, Higashidata S, Yamauchi Y (1989) J Chromatogr 461:121
30. Nitta T, Yakushijin, Kametani T, Katayama T (1990) Bull Chem Soc Jpn 63:1365
31. Gasparrini F, Misiti D, Villani C (1990) J High Resolut Chromatogr 13:182

32a. Röder W, Ruffing F-J, Schomburg G, Pirkle WH (1987) J High Resolut Chromatogr 10:665

32b. Ruffing F-J, Lux JA, Roeder W, Schomburg G (1988) Chromatographia 26:19

33. Bradshaw JS, Aggarwal SK, Rouse CJ, Tarbet BJ, Markides KE, Lee ML (1987) J Chromatogr 405:169
34. Rouse C (1988) in: Markides KE, Lee ML (eds) SFC Applications Handbook 1989. Brigham Young Univ Press, Utah, Provo, p 38
35. Frank H, Nicholson GJ, Bayer E (1978) J Chromatogr 167:187

36. Lai G, Nicholson GJ, Mühleck U, Bayer E (1991) J Chromatogr 540:217
37. Lou X, Sheng Y, Zhou L (1990) J Chromatogr 514:253
38. Lou X, Liou Y, Zhou L (1991) J Chromatogr 552:153
39. Peterson P, Markides KE, Deborah F, Johnsson DF, Rossiter BE, Bradshaw JS, Lee ML (1992) J Microcol Sep 4:155
40. Rossiter BE, Petersson P, Johnson DF, Eguchi M, Bradshaw JS, Markides KE, Lee ML (1991) Tetrahedron Letters 32:3609
41. Brügger R, Krähenbühl P, Marti A, Straub R, Arm H (1991) J Chromatogr 557:163
42. Marti A, Krähenbühl P, Brügger R, Arm H (1991) Chimia 45:13
43. Schurig V, Schmalzing D, Mühleck U, Jung M, Schleimer M, Musche P, Duvekot C, Buyten JC (1990) J High Resolut Chromatogr 13:713
44. Schurig V, Juvancz Z, Nicholson G, Schmalzing D (1991) J High Resolut Chromatogr 14:58
45. Schmalzing D, Nicholson G, Jung M, Schurig V (1992) J Microcol Sep 4:23
46. Schurig V, Schmalzing D, Schleimer M (1991) Angew Chem Int Ed Engl 8:30
47. Schleimer M, Schurig V, submitted for publication

9 Supercritical Fluid Chromatography/ Mass Spectrometry

J. DAVID PINKSTON

9.1 Introduction

The field of supercritical fluid chromatography (SFC) has experienced a phenomenal growth during the decade of the 1980s, both in terms of publications and practitioners. A noteworthy observation is that the early publications in capillary SFC [1–3] were quickly followed by descriptions of capillary SFC/mass spectrometry (SFC/MS) [4, 5]. These authors recognized the unique power of MS for the identification of chromatographic effluents. While other detectors have unique advantages, none has the combination of sensitivity, selectivity, universality, and wide compatibility possessed by the mass spectrometer.

After a brief description of early work in SFC/MS, the bulk of the chapter will be devoted to a critical analysis of the capabilities, advantages, and shortcomings of the various types of SFC/MS interfaces. The choice of column type (i.e., flow rate) often dictates which interface, which ionization methods, and even which types of mass spectrometer may or may not be used. Thus one section will treat packed-column SFC/MS, while another will treat capillary (wall coated open tubular)-column SFC/MS. The direct introduction of analytes dissolved in supercritical fluids into a mass spectrometer without on-line chromatography (supercritical fluid injection/ mass spectrometry) will be treated in the final section. While relatively few investigators have pursued this latter combination of supercritical fluids with mass spectrometry, it is an area that is bound to grow as scores of laboratories begin using analytical-scale supercritical fluid extraction.

9.2 Early Investigations of SFC/MS

Milne first proposed mass spectrometric detection for SFC [6]. He envisioned an arrangement typical for molecular beam studies, involving a nozzle/skimmer combination and multiple stages of pumping to allow electron ionization (EI). The dissociation of solute/solvent clusters in the expansion of a supercritical fluid solution into a vacuum was studied by Giddings et al. [7]. Theoretical considerations were accompanied by experimental results. For example, they studied the transport of squalene in supercritical argon (1360 atm) through a series of pressure reduction steps (consisting of an expansion nozzle, a capillary restrictor, and an effusion separator in series) into the ion source of a mass spectrometer.

Randall and Wahrhaftig provided the first report of the coupling of SFC and MS [8, 9]. They used an expansion nozzle/skimmer/collimator arrangement and multiple stages of pumping to form a molecular beam from the SFC effluent. They obtained EI spectra of aromatics introduced in supercritical ethylene, ethane, or carbon dioxide. However, the sensitivity of the combination was poor (μg quantities were required), and the laser-drilled expansion-nozzle aperture was subject to frequent plugging.

Gouw et al. reported the combination of SFC with simultaneous mass spectrometric detection and high pressure fraction collection [10]. They used a metal capillary to introduce a portion of the effluent directly into the ion source. The capillary was crimped to restrict the flow, yet it was crimped at its entrance rather than its exit. Perhaps this explains why considerable tailing was observed in chromatograms of polynuclear aromatic hydrocarbons (PAHs) despite the fact that the interface was heated to 450 °C.

9.3 SFC/MS Using Direct Introduction Interfaces

The early studies in SFC/MS discussed above all used magnetic-sector mass spectrometers and packed-column SFC. Packed-column flow rates are high (typically >20 mL/min at STP), while sector mass spectrometers are ion-beam instruments which do not tolerate high pressures ($<1\times10^{-5}$ Torr). The next generation of SFC/MS instrumentation, ushered in with the earliest publications in capillary SFC, had a different face altogether. The coupling of small internal diameter (<200 μm) fused silica capillary columns with quadrupole mass spectrometers shrank the fundamental gap that existed between a high pressure, high flow rate chromatographic technique and a moderate vacuum mass spectrometer. Capillary column flow rates are typically less than 2 mL/min at STP, and quadrupole mass spectrometers are generally more tolerant of high pressures than are sector instruments. Thus Smith, Fjeldsted, Lee, and coworkers described a much simpler SFC/MS interface than the molecular beam interface discussed above [4, 5, 11, 12]. In their "direct fluid injection" interface the capillary column was linked to a platinum-iridium capillary. The end of the metal capillary was either crimped to produce a flow restrictor, or it was butted to a stainless steel disk with a laser-drilled orifice serving as the flow restrictor (see Fig. 1). The chromatographic effluent was introduced through the flow restrictor directly into an unmodified, "simultaneous" EI/CI ion source.

The interface probe was heated to the temperature of the SFC oven. Thus the interface was relatively simple and provided good chromatographic fidelity.

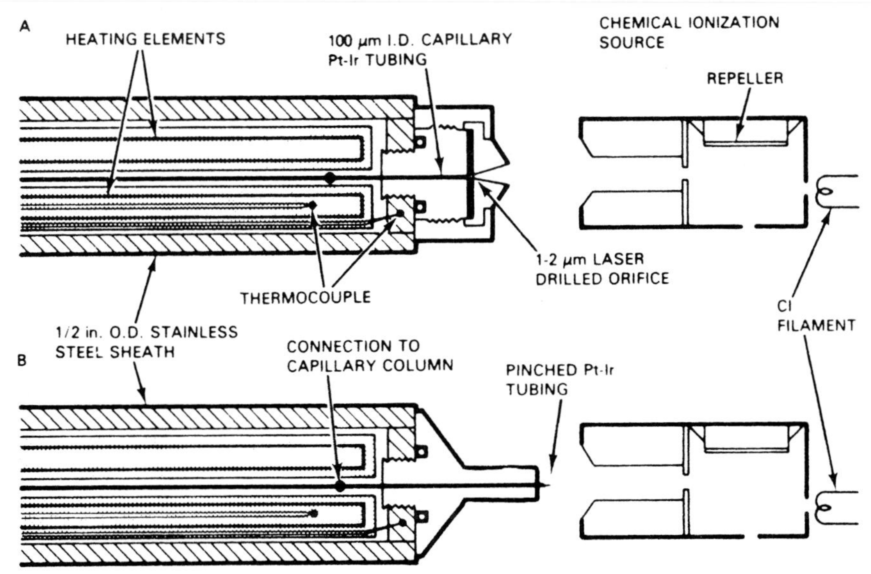

Fig. 1 A, B. Schematic diagram of two DFI probes designed by Smith et al. for interfacing SFC with a conventional CI ion source: **(A)** using a non-viscous laser-drilled orifice restrictor; **(B)** using a pinched capillary restrictor. Reproduced from Ref. [4] with permission from Elsevier Scientific Publishing Company

9.3.1 Flow Restriction

Most capillary SFC/MS has been performed using direct fluid injection type interfaces which are in many ways similar to the simple and effective interface described above. The major difference has been in the nature of the flow restrictor. Most DFI interfaces now use fused-silica-capillary-based restrictors such as those commonly used in SFC with flame ionization detection. Narrow, straight-walled capillary restrictors have been used [13, 14] successfully with volatile solutes, but they tend to plug when used with less volatile analytes. The frit [15, 16], the integral (also called the ground-tapered) [17–19], and the drawn, thin-walled tapered [20–22] restrictors have all been used with success.

It is clear that some heat must be supplied to the restrictor to counteract the Joule-Thompson cooling taking place at the point of decompression and to aid in volatilizing less volatile analytes. The optimum restrictor temperature depends upon the species being investigated. Owens et al. found that the thin-walled tapered restrictor tended to plug at temperatures of 300 °C or below during the SFC/MS separation of Triton X-100, a nonionic ethoxylated surfactant [21]. Temperatures of 350 °C and 400 °C provided statisfactory results, though the flow rate was reduced at 400 °C. Thermally-labile peroxides, on the other hand, were run at a restrictor temperature of 150 °C [23]. "Spiking" and erratic signal level were observed below 150 °C,

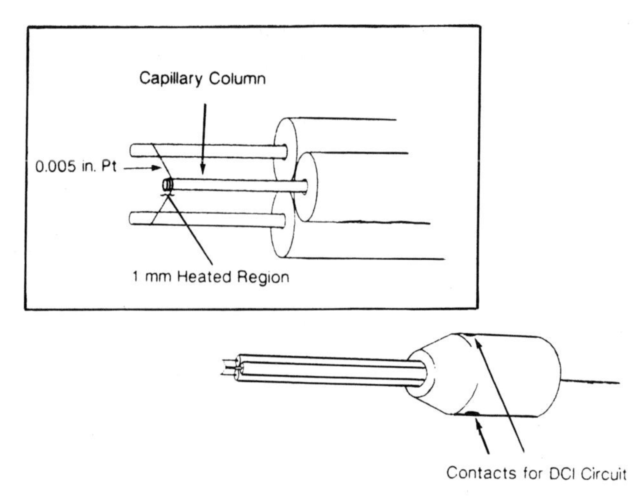

Fig. 2. Diagram of the SFC/MS probe designed by Reinhold et al. for a double-focussing sector mass spectrometer. A standard desorption/chemical ionization probe tip is configured with an additional ceramic feedthrough for the SFC column. Contact points provide electrical coupling to tabs on the side of the ion block for temperature control. The enlarged tip shows the column feedthrough with the capillary and the resistance wire for tip heating. Reproduced with permission from the American Chemical Society from Ref. [24]

while thermal degradation products increased above 150 °C. While most interface designs heat a minimum of the last 1 to 2 centimeters of the restrictor, Reinhold et al. have used a modified desorption/chemical ionization probe as an SFC/MS interface where only the last millimeter of the restrictor was heated (see Fig. 2) [24]. They found that the restrictor tended to plug at temperatures below 280 °C (estimated) while it operated well over a broad range of temperatures above this value.

9.3.2 Ionization Methods

9.3.2.1 Chemical Ionization

The flow rates used in most capillary column SFC/MS with the DFI interface are such that traditional chemical ionization (CI) is possible. Detection limits in SFC/CIMS are generally in the low to mid pg range, thus on the same order as those reported for GC/CIMS [22, 25]. Unmodified CO_2, the most common mobile phase for capillary SFC/MS, has a low proton affinity and has little effect on the CI spectra generated by common reagent gases such as CH_4, NH_3, and isobutane. Other supercritical mobile phases, such as pentane and ammonia, actively participate in CI chemistry [4, 11]. The influence of organic "modifiers" in CO_2 on chemical ionization will be discussed in "SFC/MS Using High Flow Rate Interfaces" (Sect. 4).

Concern does exist that CO_2 charge-exchange begins to compete with CI at high CO_2 flow rates [26]. This was postulated to be the cause of increased fragmentation and decreased signal-to-noise ratio (S/N) in the latter stages of a capillary SFC/MS run of ethoxylated surfactants. Chloride-attachment negative CI using difluorodichloromethane as reagent gas alleviated the problem. This drop in spectral quality and S/N has been seen by other investigators [21, 27]. Spectral quality and S/N improved upon use of a simple cryopump to reduce the pressure within the ion source manifold [21]. It was postulated that the ill-effects of high CO_2 flow rates were due to collisional processes [21, 27].

Negative CI using CO_2 as the moderating gas has been shown to yield high sensitivities with electron-capturing analytes. Electronegative groups can also be attached to analytes for improved sensitivity using NCI. Sheeley and Reinhold have added the pentafluorobenzyl ester of *p*-amino benzoic acid to the reducing end of oligosaccharides for enhanced sensitivity in SFC/NCIMS [28]. Sim and Elson have shown that certain PAH isomers that give nearly identical EI and positive CI spectra give dramatically different spectra under SFC/NCIMS conditions [29]. Roach et al. investigated the effect of CO_2 mobile phase on the electron capture, proton abstraction, and chloride attachment NCI behavior of trichothecene mycotoxins [30]. They found that CO_2 had little effect on NCI conditions.

9.3.2.2 Electron Ionization and Charge Exchange

Achieving electron ionization (EI) in SFC/MS is a worthy goal since great libraries of standard EI spectra have been compiled, and since EI fragmentation patterns can aid in structure elucidation. EI-type spectra are obtained in SFC/MS when CO_2 is used as the mobile phase with a standard EI source. However, sensitivities have been uniformly lower in SFC/EIMS than in SFC/CIMS or in GC/EIMS [31, 32]. Ion source pressure is higher in SFC/EIMS than in GC/EIMS, thus some loss in sensitivity due to decreased penetration of the ionizing electrons into the ion source has been postulated, even when a more open source modification has been used [18, 32]. Improved sensitivity has been obtained by increasing the ability of the ionizing electrons to penetrate into the ion source (i.e., increasing the electron energy) [32].

Carbon dioxide pressures in the ion source are such that concurrent EI and CO_2 charge exchange (CE) are thought to occur under the conditions of most SFC/EIMS [18, 27, 32, 33]. CO_2 CE spectra are usually very similar to EI spectra and give good EI-library matches. Lee et al. have gone one step further and increased the sensitivity and the degree of fragmentation in SFC/EIMS by the addition of more traditional CE gases such as helium [33]. Helium has a higher ionization potential than CO_2 and produces more fragmentation of typical organic compounds in CE.

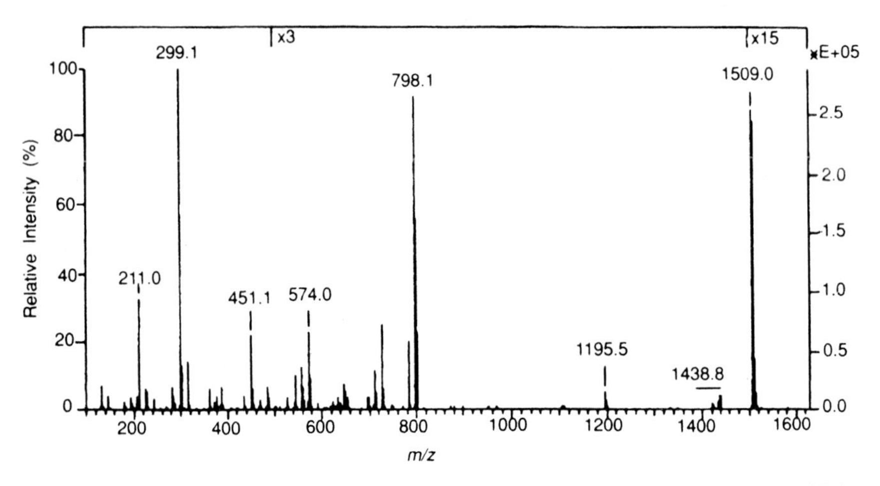

Fig. 3. EI spectrum of silylated phytic acid. Reproduced with permission from Aster Publishing Corporation from Ref. [34]

For certain less volatile and/or thermally labile materials, such as trimethylsilyl (TMS)-derivatized phosphorylated inositols [23, 34] (see Fig. 3), and *tert*-butyldimethylsilyl-derivatized poly(acrylic acid) [35], SFC/EIMS may be the best (or only) method for obtaining EI-type data. Many of these compounds are not in the EI libraries, but they do yield matches with similar compounds when library-searched. For example, the TMS derivative of inositol triphosphate is not in the EPA/NIST library but it matched well with the EI spectrum of the corresponding derivative of inositol diphosphate [23].

9.3.2.3 Secondary Ion Mass Spectrometry

Wenclawiak et al. have taken advantage of the low flow rates of capillary SFC to directly couple SFC with an unmodified secondary ion mass spectrometry (SIMS) instrument [36]. An interface probe containing a tapered capillary flow restrictor was brought near the SIMS target surface. The SFC effluent was directed at the surface and simple mixtures of low molecular weight polynuclear aromatic hydrocarbons and of phthalate esters were deposited on the surface. Secondary ions related to the analytes were then desorbed from the surface using a 3 keV Xe^+ primary ion beam. Instrumental limitations such as scan rate prevented application of this SFC/SIMS instrument to more complex mixtures. Yet the authors suggest a number of improvements, such as cleansing of the target surface between chromatographic peaks by sputtering, the use of a revolving, multiple-target-surface system, and faster scan rates to improve the analytical capabilities of SFC/SIMS. Off-line SFC/SIMS will be discussed in a later section.

9.3.3 Type of Mass Analyzer

9.3.3.1 DFI SFC/MS Using Quadrupole Mass Spectrometers

Most DFI has been performed using quadrupole instruments. Quadrupole mass spectrometer ion sources operate at or near ground potential, are tolerant of high pressures (ion transmission level holds through the 10^{-5} Torr range), are easily computer controlled, and are relatively inexpensive. All these advantages make them attractive for DFI SFC/MS. Most manufacturers now offer quadrupole instruments with extended mass ranges (upper m/z limit typically 4000), thus bringing the capabilities of quads more in line with those of SFC [22]. In general, quadrupole mass spectrometers used for DFI SFC/MS have been used without modification, with the exception of the more-open ion source design [18] and the cryopumping experiments [21] discussed earlier. Many investigators have striven to design systems which can be rapidly converted from SFC/MS to GC/MS operation. Some have pushed an integral or frit restrictor up to the ion source through an unmodified "capillary direct" GC/MS interface block [17, 19, 20, 37]. Simply removing the restrictor and inserting the GC capillary column readies the instrument for GC/MS. Other workers have designed interface probes, most often introduced through the direct introduction probe vacuum lock, which can be either left in the instrument or removed during GC/MS operation [4, 21, 22, 29, 38–40].

Lee et al. have used an inexpensive benchtop quadrupole mass spectrometer for DFI SFC/MS. The only significant modification in their first report was the use of a chemical ionization source [20]. The authors acknowledged that additional pumping would provide enhanced performance. This step was implemented and described in a second publication [33]. Despite the additional pumping, conventional EI SFC/MS required an order-of-magnitude more material than He CE SFC/MS or EI GC/MS to obtain a representative spectrum. One major limitation of such benchtop quadrupole instruments of SFC/MS is their limited mass range, with upper m/z limits typically below 1000.

Quadrupole instruments are not able to perform high resolution exact mass measurements at part-per-million accuracies, an important tool in structure elucidation and targeted compound analysis. Also, the practical mass range of quadrupole instruments, in which reasonable mass resolution can be maintained with good sensitivity, is still limited by both electronics and machining, despite the advances described above. Thus many laboratories have investigated the use of other types of mass spectrometers, which outperform quads in these areas, for DFI SFC/MS.

9.3.3.2 DFI SFC/MS Using Double-Focussing Sector Mass Spectrometers

Double-focussing (dual electric/magnetic) sector instruments present certain advantages over quadrupoles. The higher resolving power of sector instruments enables them to perform accurate mass and relative intensity measurements on an ion in the presence of other ions of the same nominal mass. The practical mass range of high performance sector instruments is typically higher (10000 Da or even beyond at full accelerating voltage) than that of quadrupole instruments. Yet performing DFI SFC/MS on a sector instrument is not without certain complicating factors. Sector ion sources typically operate at high voltage (up to 8 kV or even higher). A DFI interface must be designed to prevent high voltage discharge. Such a probe was designed by Kalinoski et al. and is shown in Fig. 4 [41]. Reinhold et al. have modified a desorption/chemical ionization probe for DFI SFC/MS as discussed earlier (see Fig. 2) [24]. Their work is a good example of the use of the high mass range of double-focussing instruments to more adequately cover the molecular weight range of SFC. Figure 5 shows the ammonia CI SFC/MS spectrum of a derivatized glucose oligomer containing 15 glucose units. Figure 6 shows mass chromatograms of the ammonium adduct ions of a series of derivatized glucose oligomers.

The electric and magnetic sectors of a sector mass spectrometer disperse beams of ions according to energy and momentum, respectively, to achieve high mass resolution. Since sector mass analyzers are ion beam instruments, they are not as tolerant of high pressures as are quadrupole analyzers. For this reason denser frit restrictors were used by Huang et al. to reduce the flow of mobile phase entering the mass spectrometer in their DFI interface designs [15, 16].

9.3.3.3 DFI SFC/MS Using Fourier Transform Mass Spectrometers

Fourier transform mass spectrometry (FTMS) has the potential for ultrahigh mass resolving power (10^6 in the m/z 100 range) if pressure within the mass analysis cell is low ($<10^{-8}$ Torr). It also has the potential for high mass range. Excellent reviews have appeared which describe FTMS and its use for chromatographic (especially GC) detection [42–44]. Two reports on the use of FTMS for DFI SFC/MS have appeared [45, 46]. Both used the dual-cell FTMS design to minimize the adverse effects of high pressure on performance. The chromatographic effluent is introduced and ions are created in one cell. The ions are then transferred to a second, low pressure cell where mass analysis takes place.

Lee at al. were able to obtain an exact mass measurement of the molecular ion of caffeine (m/z 194) within 0.036 ppm of the calculated value [45]. Despite relatively high pressures (10^{-7} torr range) in the analysis cell, they were able to obtain a mass resolution of 8300 for the molecular ion of

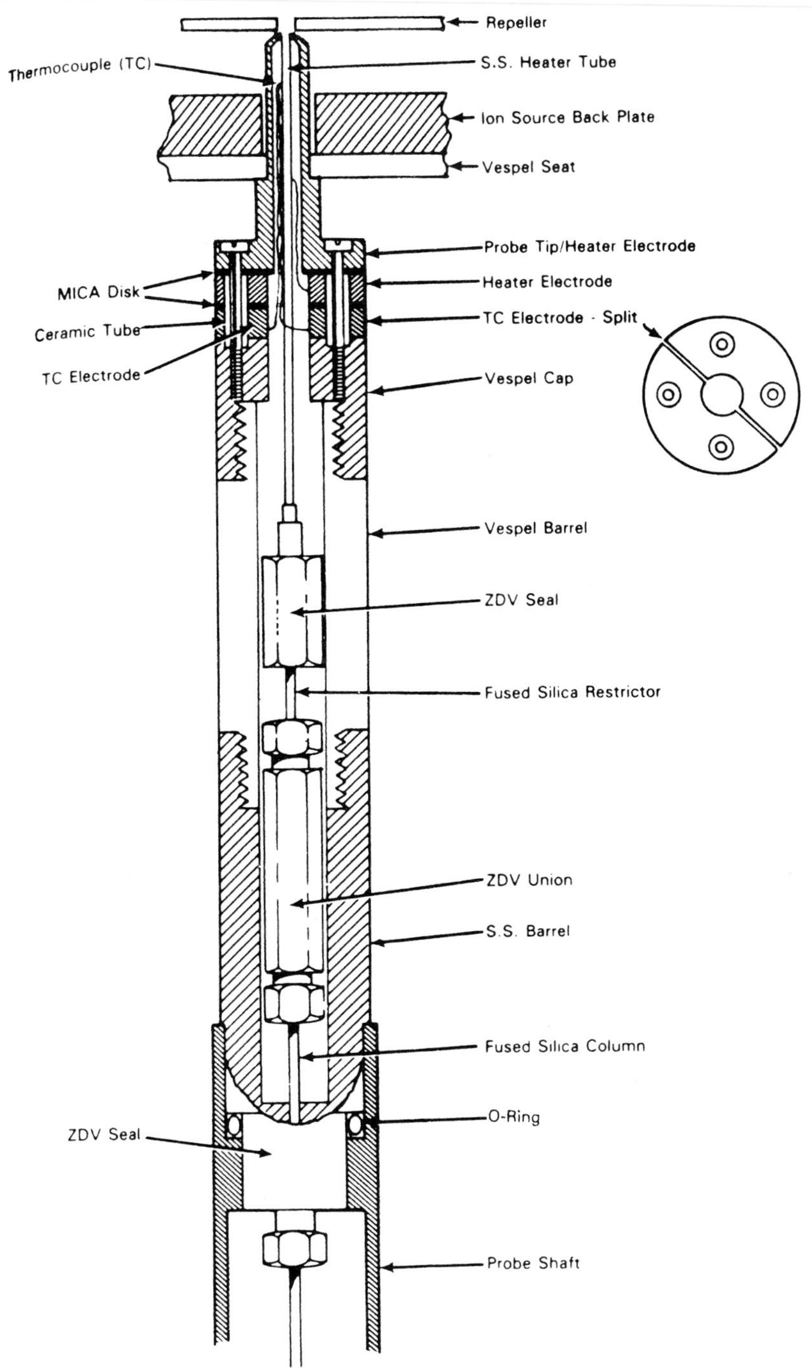

Fig. 4. Schematic diagram of the SFC/MS interface probe designed by Kalinoski et al. for the VG ZAB (high voltage) CI source. Reproduced from Ref. [41] with permission from Elsevier Scientific Publishing Company

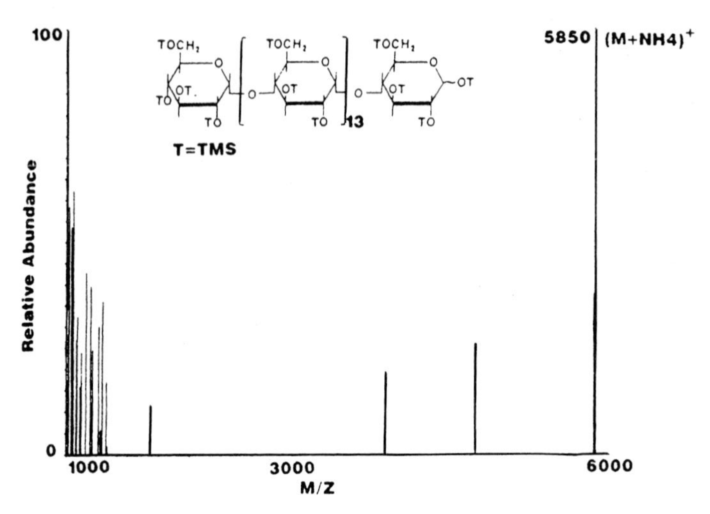

Fig. 5. Ammonia CI mass spectrum of trimethylsilylated glucose oligomer with a degree of polymerization of 15. Reproduced from Ref. [24] with permission from the American Chemical Society

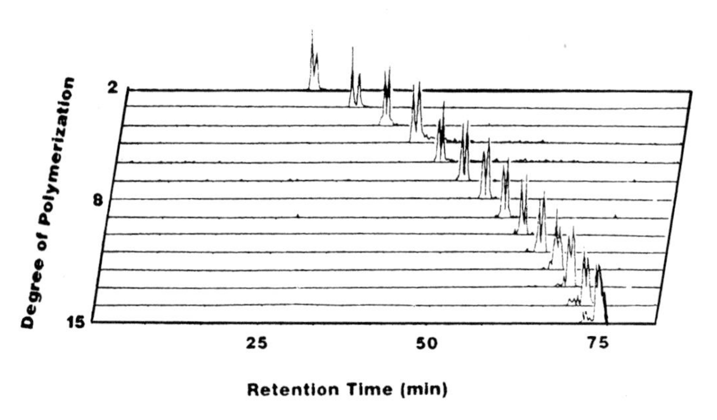

Fig. 6. Mass chromatograms of the ammonium adduct ions of a mixture of trimethylsilylated glucose polymers. Reproduced from Ref. [24] with permission from the American Chemical Society

caffeine. Both EI and CI spectra were obtained. Unfortunately, it was necessary to use a 1-m long transfer line to introduce the chromatographic effluent into the cell, and the tip of the flow restrictor was not heated above 100 °C. Thus it was not possible to introduce higher molecular weight, less volatile solutes to the FTMS instrument. Laude et al. [46] obtained similar results, though they used a 100-μm i.d. SFC column rather than a 50-μm i.d. column as used by Lee et al. Laude et al. did use a heated restrictor and analyzed a wider variety of mixtures, including polynuclear aromatic hydrocarbons, barbiturates, and pesticides. In addition to investigating the pressure limits on mass resolution and spectral S/N in SFC/FTMS, Laude et al. attempted to deal with the dynamic range limitations in FTMS. Dynamic range is adversely affected by ion space-charge effects in FTMS. They reduced the electron energy from 70 eV to 13 eV to minimize the ionization of CO_2 (ionization potential 13.8 eV), and continuously ejected any ions formed from CO_2. Even these measures did not alleviate the space-charge problems at high SFC pressures. They concluded that lower pressures would be required for significant progress.

9.3.3.4 DFI SFC/MS Using Quadrupole Ion Trap Mass Spectrometers

The quadrupole ion trap has been used for trapped-ion studies for many years. A relatively inexpensive, commercial version of the ion trap, the ion trap detector (ITD), has been introduced for routine capillary GC/MS. It operates on the principle of mass selective instability. Ions are generated inside the trap by either EI or CI and are stored. In its simplest implementation, the trapping parameters are scanned such that ions of successively larger m/z no longer have stable trajectories within the z axis of the trap. As they exit the trap some of these ions collide with an electron multiplier and are detected. Stafford et al. found that operating the trap with a relatively high pressure of a gas such as He (10^3 Torr) significantly improved mass resolution and S/N [47]. This is presumably due to collisional damping of the ions' motion, causing them to move to the center of the trap from where they are more efficiently ejected and detected. The mass range of the commercial ITD is relatively low (<1000 Da) but efforts to increase the mass range by changing the size and operating voltages of the trap have been successful. Hemberger et al. presented spectra extending to m/z 45000 [48].

Given the potential for extended mass range, the ITD would seem to be a promising, relatively inexpensive mass analyzer for SFC/MS. Todd et al. recently modified an ITD by removing the Teflon insulating rings located between the ring and end-cap electrodes to increase the pumping speed [49]. They studied the ion trap pressure at various mobile phase pressures and restrictor temperatures. They obtained EI-like spectra of aromatics which provided good matches with the NIST library spectra. Intensities were approximately seven-fold greater when He was added to the CO_2 effluent in the trap. Yet S/N ratios were disappointing. The authors concluded that additional pumping would probably be required for further progress.

9.4 SFC/MS Using High-Flow-Rate Interfaces

Packed-column SFC has grown in popularity due to speed of analysis and freedom from many of the injection, detection, and dead-volume constraints of capillary-column SFC. Packed-column SFC using unmodified CO_2 is well suited to the analysis of relatively nonpolar mixtures, while more polar species can be eluted by adding small amounts of polar "modifiers", such as low molecular weight alcohols or water. Research is proceeding in many laboratories with the goal of producing less active or better deactivated packings for packed-column SFC. Other researchers have advocated the use of capillary columns with high flow rates of supercritical mobile phase [25].

The mobile phase flow rates of both packed SFC and high-flow-rate capillary SFC are too high for the entire effluent to be directly injected into the ion source of a traditional mass spectrometer. Those developing or applying packed or high-flow-rate capillary SFC/MS have used a variety of interfaces. Some simply serve to split the effluent, while others reduce the total flow rate into the mass spectrometer while enriching the flow in the species of interest. The following sections will review the advantages and drawbacks of the various high-flow-rate interfaces. The effects of mobile phase modifiers on the characteristics of each interface will be discussed since modifiers are so often used in packed SFC.

9.4.1 SFC/MS Using the Moving Belt Interface

The moving belt interface was originally developed for LC/MS [50]. In its most widely successful form the chromatographic effluent is spray-deposited on a stainless steel or Kapton belt. The belt then moves though a series of vacuum locks where the solvent evaporates, sometimes aided by heat. The analytes left on the belt are then desorbed by rapid heating as the belt nears the ion source. The belt moves through a "clean up" heater on its way back to the spray deposition area. One significant advantage of the moving belt interface is that true EI and all modes of CI are available, since the solvent is no longer present when the analytes enter the ion source.

The group led by Games has made extensive use of the moving belt interface for packed SFC/MS [51–55]. The mobile phase most often used was CO_2 with a small amount of a polar modifier such as a low-molecular-weight alcohol. The moving belt interface is even better suited to packed SFC than to LC, since most of the SFC mobile phase is already a gas upon decompression. In fact, Berry, Games, and Perkins found that this decompression forced a modification to their original belt-deposition device [51, 52]. The original, unheated, stainless steel line linking the moving belt to the chromatograph was replaced by a Finnigan-MAT thermospray deposition device because of ice build-up on the tip. The end of the thermospray deposition device was crimped to provide a restriction and was heated to compensate for cooling due to the decompression. They illustrated the versatility of the moving belt interface with both EI and CI SFC/MS runs of xanthines, sulfonamide drugs, and an ergot extract using a quadrupole instrument [52]. True EI spectra were obtained despite the use of methanol and methoxyethanol as modifiers.

In later work, Balsevich et al. used the moving belt interface to couple packed SFC and a sector mass spectrometer [54]. They obtained EI spectra of indole alkaloids from leaves of *Catharanthus roseus*. The spectra provided good matches to standard EI spectra despite the use of a 5% – 15% methanol modifier gradient. Ramsey et al. reported the use of the moving belt interface to couple packed SFC and a sector-quadrupole hybrid mass spectrometer [55]. They used on-line SFE/packed SFC/MS/MS for the extraction, separation, and identification of veterinary drug residues in pig's kidney. Niessen et al. have also used the moving belt interface to obtain true EI in the packed SFC/MS analysis of the herbicide diuron in plasma [56]. Diuron is not degraded under the conditions of the analysis, though diuron is susceptible to thermal degradation.

The moving belt interface also has its drawbacks. Fairly rigid volatility limits are imposed on the analytes. Workers have observed both the loss of volatile analytes [51], and thermal degradation of less volatile ones [53]. The moving belt interface is also mechanically complex.

9.4.2 SFC/MS Using the Particle Beam Interface

The particle beam interface, with similar implementations known as the momentum separator and the monodisperse aerosol generation interface, is another enrichment interface. Like the moving belt interface, the particle beam interface was originally designed for LC/MS. Formation of the particle beam first begins with nebulization of the chromatographic effluent using a nebulizing gas such as He. The HPLC mobile phase begins to evaporate as the effluent droplets and the nebulizing gas move through a desolvation region toward the baffles of the particle beam interface. Heat is generally provided to the desolvation chamber to compensate for cooling due to solvent evaporation. The particle beam interface baffles are arranged in a nozzle/skimmer fashion, much like an oversized jet separator for GC/MS. As the solute particles pass through the baffles of the interface, mechanical pumps remove most of the evaporated solvent and nebulizing gas. The particles finally move into the ion source, impact on its heated walls, and are vaporized. Since the bulk of the mobile phase is not present in the ion source, both EI and all types of conventional CI are available. The particle beam interface can accept high flow rates of HPLC mobile phases (up to 2 mL/min in favorable cases).

The group led by Henion has investigated the use of the particle beam interface for packed-column SFC/MS [57, 58]. They used a linear, fused silica flow restrictor which was housed within a modified thermospray probe and was heated to ca. 60 °C. The He nebulizing gas was introduced coaxially to the flow restrictor. They applied the particle beam SFC/MS interface to a variety of mixtures, such as herbicides, pesticides, polymer additives, steroids, and veterinary drugs. Henion and coworkers expected the particles from the supercritical fluid expansion to be smaller than those obtained during the evaporation of liquid effluents in particle beam LC/MS. They varied the spacings and sizes of the skimmers in an attempt to optimize SFC/MS performance [58]. However, the quantities of the types of compounds listed above required to obtain a representative EI spectrum were universally high, in the low microgram range. They found that proper positioning of the restrictor tip and adjustment of the nebulizer gas were critical in assuring reliable performance. The particle beam interface should suffer from some of the same analyte-volatility limitations as the moving belt interface. Analytes with high volatilities will be lost in the particle beam interface. Certain analytes will have volatilities too low for vaporization when the particles strike the hot walls of the ion source. Other analytes may undergo thermal degradation before they are volatilized. Despite these limitations and the low responses observed, Henion and coworkers found the particle beam interface to be rugged and widely applicable.

9.4.3 Off-line SFC/MS

Perhaps the ultimate enrichment "interface" is off-line coupling of SFC with MS, where the mobile phase is removed entirely from the analyte. This type of approach is simple and effective when a few peaks in an SFC run require mass spectrometric identification but the mass spectrometer at hand is not equipped with an on-line interface. For example, the nature of plasma desorption mass spectrometry (PDMS) precludes its on-line coupling with chromatographic techniques. Keough and Pinkston have used off-line SFC/PDMS to identify key peaks in SFC runs of nonionic surfactants [59]. The capillary SFC separation was monitored by UV detection. When a peak of interest eluted, a PDMS sample foil was held near the tip of the heated, tapered-capillary flow restrictor. The peak of interest was collected on the foil and the foil was later subjected to PDMS analysis. The nonionic surfactants produced extremely intense PDMS spectra. The CO_2 mobile phase permits this simple but effective fraction collection.

Wenclawiak advocated off-line coupling for SFC/secondary ion mass spectrometry (SIMS) [60]. A mixture of steroids was first separated by SFC/FID. The FID was then replaced by a home-built fraction collector containing a silver target surface. Using the retention times from the first run, a fraction containing two closely-eluting steroids was collected on the target during a second run. The target was then subjected to SIMS analysis. Identifications were aided by silver cationization of the species of interest.

9.4.4 SFC/MS Using Direct Introduction of Split Effluents

Perhaps the simplest method of interfacing a high-flow-rate SFC technique to MS is to direct only a portion of the chromatographic effluent into the mass spectrometer's ion source. This is entirely reasonable when sample quantities are not limited. In addition, the effluent not directed toward the ion source can be fed to another detector, such as an FID, an FTIR, or an element-selective detector, in order to obtain additional information about the sample. The most obvious implementation of this technique is to split the effluent before it enters the ion source. This approach will be discussed in this section. The effluent may also be split after it enters the mass spectrometer but before ionization (i.e., a portion of the effluent is pumped away before entering the ionization region), or the effluent (and ions) may even be split after the ionization step. These latter approaches will be discussed in subsequent sections. Unlike the enrichment and solvent elimination interfaces, the mobile phase usually plays a direct role in ionization in splitting interfaces. For example, methanol CI is obtained when methanol-modified CO_2 is used as the mobile phase.

Gouw et al. used effluent splitting to simultaneously interface packed-column SFC to a magnetic sector mass spectrometer and a high pressure fraction collector [10]. As discussed in Sect. 2, a crimped metal capillary located within a "tee" at the outlet of the chromatographic column was used to introduce a portion of the effluent directly into the ion source. The balance of the effluent was directed toward the fraction-collection device.

Crowther and Henion used a modified commercial "direct liquid introduction" (DLI) probe originally designed for LC/MS [61]. The instrument was also equipped with a liquid nitrogen cryopump. Approximately 1/12 of the effluent (0.8 to 1.5 mL/min) from the 2.1-mm i.d., 20-cm long packed column was introduced to the CI ion source through a laser-drilled pinhole orifice. The methanol mobile phase modifier doubled as a mild chemical ionization reagent gas. The authors investigated the ionization and elution characteristics of pharmacological alkaloids and veterinary drug standards. Crowther and Henion found significant advantages over direct liquid introduction LC/MS. However, the detection limits of DLI SFC/MS were still in the mid-nanogram range (due to effluent splitting at the interface), and the pinhole diaphragms had a relatively short lifetime (3 to 10 days).

9.4.5 SFC/MS Using Post-Expansion Splitting

A number of investigators have coupled high-flow-rate SFC techniques to MS by one or more stages of pumping between the region of mobile-phase expansion (i.e., the tip of the flow restrictor) and the ionization region. In general, these interfaces are not designed to enrich the flow entering the ion source in analyte. This approach to high-flow-rate SFC/MS is defined here as post-expansion splitting. Though this is quite different in appearance from directing only a fraction of the effluent to the SFC/MS interface, as discussed above, both methods, in effect, split the effluent to bring the chromatographic flow rate to a level compatible with traditional ionization methods. Since a fraction of the mobile phase enters the ion source along with a fraction of the analyte, mobile phase modifiers may have a profound impact on the ionization mechanism.

Randall and Wahrhaftig's early work is an example of post-expansion splitting [8, 9]. They used multiple stages of pumping to bring the influx of analyte and mobile phase into the ion source down to a level compatible with EI. As discussed earlier, this system had low response. Much of the analyte was probably pumped away along with the mobile phase within the stages of pumping.

The interface designed by Matsumoto et al. [62] used both effluent splitting and post-expansion splitting. Hexane and ethanol-modified hexane were used as mobile phases for the 500-μm-i.d., 70-cm-long packed capillary column. Approximately half of the effluent (40 μL/min) was introduced to the nebulizing interface through a 12-μm-i.d. capillary flow restrictor. Nebulization was assisted by a coaxial flow of He gas saturated with methanol. A second form of splitting took place within the heated nebulization region: much of the expanding effluent and nebulizing gas was removed by a mechanical pump. The remaining mixture reached the ion source where ethanol/methanol CI took place. Application to a wide range of mixtures,

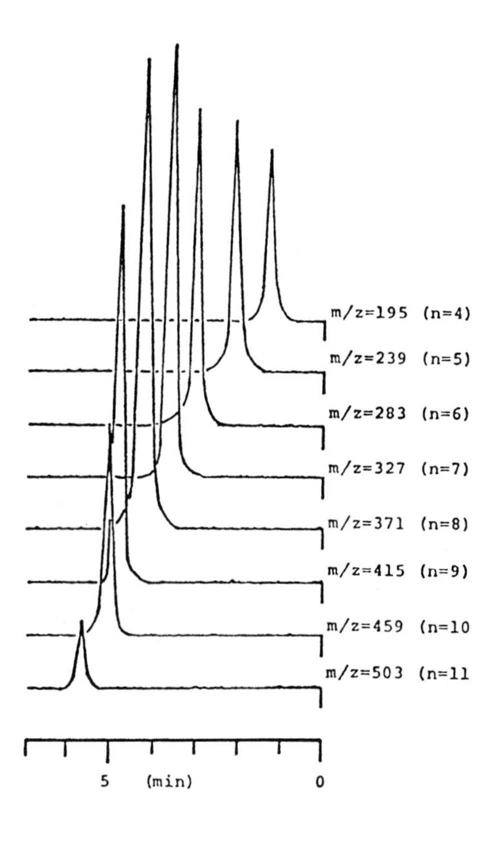

Fig. 7. Mass chromatograms of the protonated molecules of the components of PEG-400. Reproduced with permission from Nippon Shitsuryo Bunseki Gakki from Ref. [64]

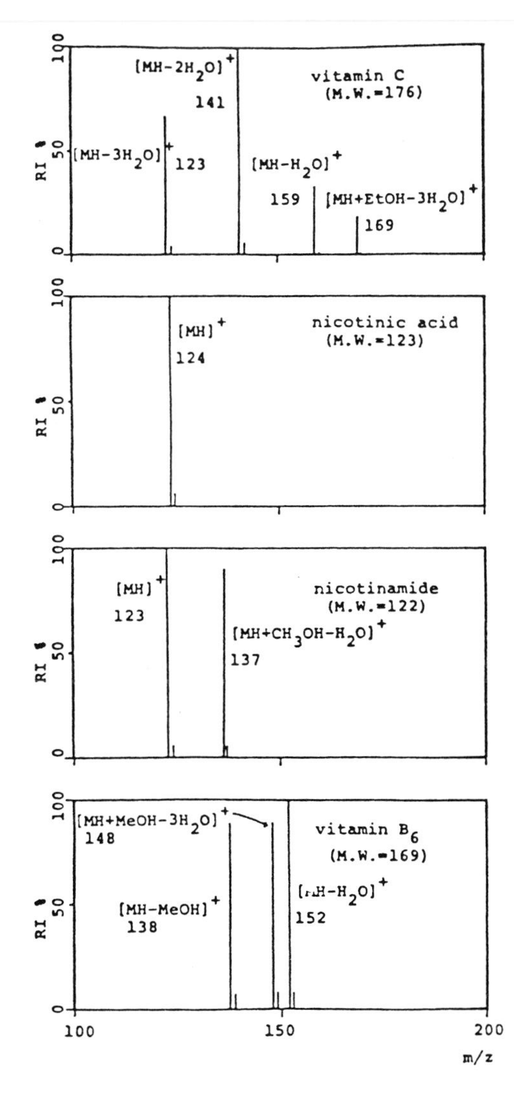

Fig. 8. Mass spectra of some water-soluble vitamins. Reproduced with permission from Vieweg Verlagsgesellschaft from Ref. [63]

including polystyrene [62], triglycerides [62], phthalate esters [62, 63], non-ionic surfactants [62, 64], and both fat-soluble and water-soluble vitamins [63] were demonstrated, though the upper mass limit of the mass spectrometer was only 550 Da. Figure 7 shows the mass chromatograms of the protonated molecules of the components of poly(ethylene glycol) 400. Figure 8 shows mass spectra of four water-soluble vitamins. The mobile phase used to elute these polar species from the packed capillary was 1 : 1 hexane to ethanol at 260 °C.

Smith and Udseth designed an interface for microbore (packed) and high-flow-rate capillary SFC/MS which used post-expansion splitting [25].

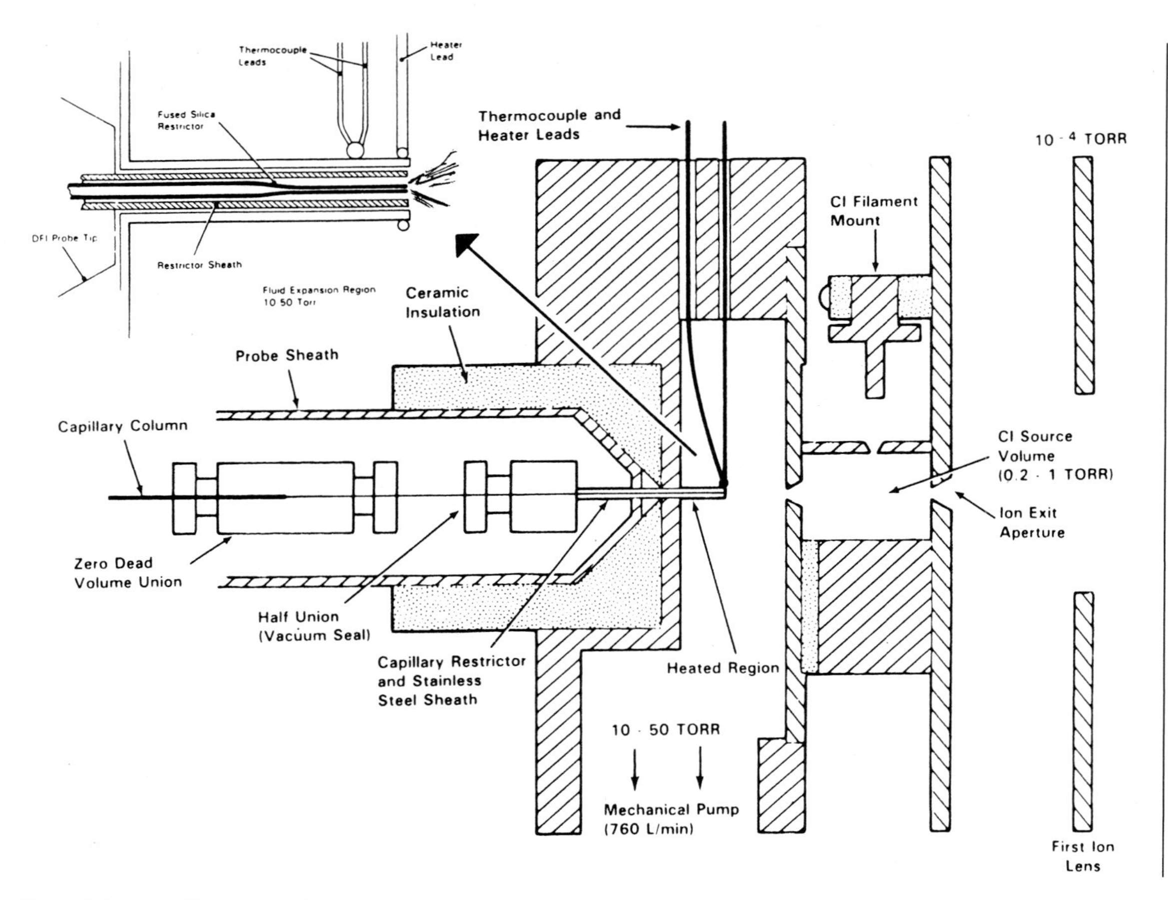

Fig. 9. Schematic illustration of the high-flow-rate SFC/MS interface designed by Smith and coworkers. The inset shows details of the probe tip assembly and expansion region. Reproduced from Ref. [66] with permission from Preston Publica-tions

Figure 9 shows the design. The column was linked to a tapered capillary flow restrictor which was housed in a stainless steel capillary sheath for mechanical support. The sheath was part of the interface probe and was itself inserted into a larger, heated stainless steel capillary which was part of the high-flow-rate (HFR) interface. The expansion region was pumped by a mechanical pump. The size of CI source entrance aperture was chosen to provide the desired split between the effluent going to the pump and that going to the CI source (a typical value was 25 : 1).

The authors claimed an enhancement in sensitivity vs previous interfaces, especially for higher molecular weight components. They suggested that this enhancement might have been related to the alignment of the restrictor with the CI entrance aperture, an effect reminiscent of the action of the particle beam interface discussed earlier. Other factors related to this enhancement were also suggested, such as more ideal expansion (delayed solute nucleation) due to the larger diameter restrictor aperture, optimized CI conditions, and injection of analytes into a region of the CI source from which ions are more efficiently sampled.

Smith and coworkers used the HFR interface for a variety of applications [65–67]. Kalinoski and Smith used microbore SFC/MS for the characterization of mixtures of thermally labile organophosphorus insecticides [65]. Ninety-four picograms of chlorpyrifos gave a S/N value of 21 in the selected

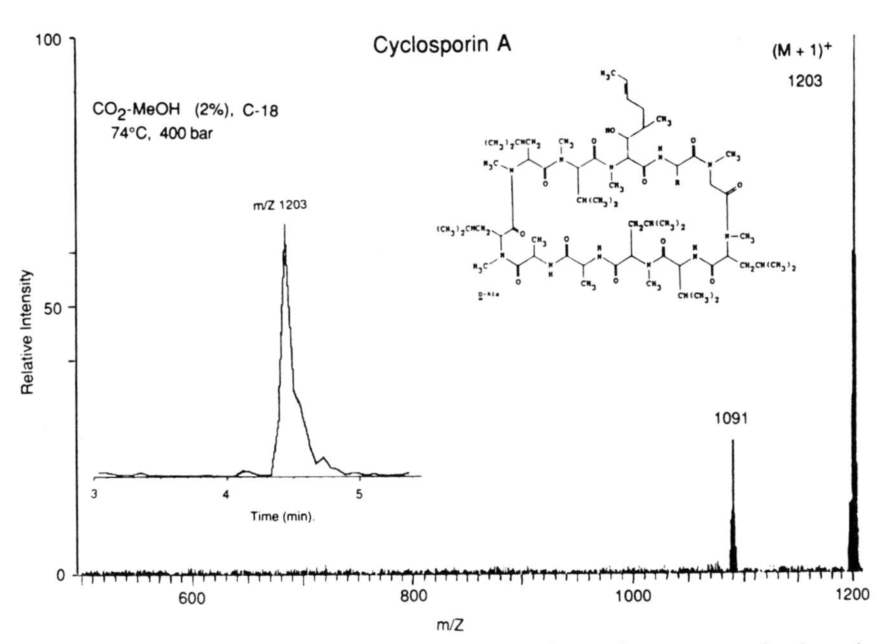

Fig. 10. Positive ion methanol CI spectrum, structure, and mass chromatogram of cyclosporin A obtained by packed SFC/MS using the high-flow-rate designed by Smith and coworkers. Reproduced from Ref. [67] with permission from John Wiley and Sons

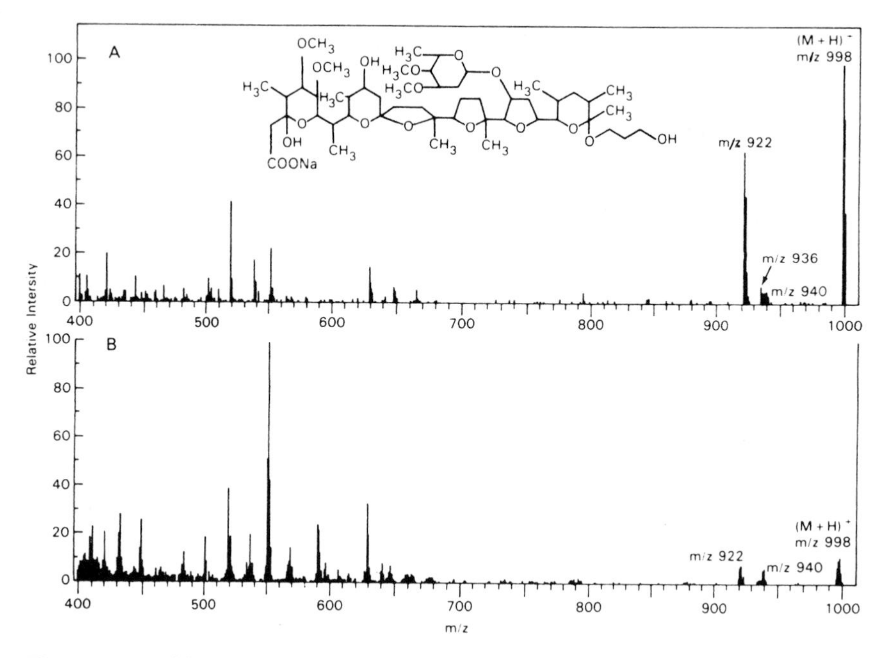

Fig. 11 A, B. Positive ion 2-propanol CI spectrum of a thermally-labile polyether antibiotic ionophore (inset) obtained by capillary SFC/MC using the high-flow-rate interface designed by Smith and coworkers at low (**A**) and high (**B**) restrictor-heater temperatures. Reproduced from Ref. [67] with permission from John Wiley and Sons

ion monitoring mode. Wright et al. used high flow rate capillary and microbore SFC/MS to study complex mixtures of hydrocarbons derived from both petroleum and coal [66]. Wright and Smith used a variety of mobile phases, including pentane, 2-propanol modified pentane, 2-propanol modified propane, and 2-propanol modified CO_2, for SFC/UV and SFC/MS of porphyrins [68].

Especially impressive were the SFC/MS spectra obtained by Kalinoski et al. of the cyclic undecapeptide cyclosporin A, shown in Fig. 10, and of two ionophores (ionic polyether antibiotics) using the high flow rate interface [67]. The methanol or 2-propanol mobile phase modifiers were used as CI reagents. Figure 11 shows the spectra of one of the ionophores collected at a restrictor temperature of 145 °C (top) and at 300 °C (bottom). The spectra clearly illustrate the importance of maintaining a low restrictor temperature when analyzing thermally-labile compounds by SFC/MS.

9.4.6 SFC/MS Using the Thermospray Interface

The thermospray interface was originally designed for coupling analytical scale HPLC to mass spectrometry [69]. Nebulization of the effluent is assisted by heat. Ions in solution may be expelled from the droplet as the solvent evaporates and the droplet diameter decreases. Ionization may also be assisted by discharge or high energy electrons. In most designs a "repeller" electrode directs ions toward a small aperture leading to the higher vacuum region of the mass spectrometer. Only a small fraction of the ions produced actually penetrate the aperture and are mass analyzed. The bulk of the ions and neutrals are pumped from the thermospray source by a mechanical pump. In keeping with the previous two sections, one might label this interface the "post-ionization splitting" interface, as suggested by Chapman [70].

The group led by Games has actively pursued the interfacing of packed SFC with mass spectrometry using a modified thermospray source [53, 54, 71, 72]. They were motivated in this work by thermal degradation observed using the moving-belt interface [53, 71]. The modification consisted of replacing the coiled "vaporizer" tube with a straight segment of stainless steel capillary tubing which was crimped at its outlet to provide a restriction in flow. Typical packed-column flow rates were 4 mL/min, of which approximately half was directed to the thermospray interface.

In a comparison using a crude ergot fermentation extract as sample, the thermospray interface provided better sensitivity than the moving-belt interface. A number of compounds not seen with the moving-belt interface were detected with the thermospray interface [71]. Along with this advantage arose a potential disadvantage of the thermospray interface: as with the other splitting interfaces, the mobile phase is present during ionization and thus may play a major role in the ionization. For example, the ergot fermentation extract was run with methanol-modified CO_2 and high-energy-electron ionization ("filament-on mode"). The resulting ionization mode was methanol CI, which provided spectra with less structural information than those obtained with the moving-belt interface. On the other hand, EI type spectra were obtained when high-energy-electron ionization was used with unmodified CO_2, presumably due to charge-exchange ionization.

Similar results to those obtained with the ergot fermentation extract were obtained by moving-belt and thermospray SFC/MS of indole alkaloids from *Catharanthus roseus* by Balsevich et al. [54]. Approximately 60 alkaloids were detected using filament-on thermospray. Fourteen alkaloids were identified by matching high-quality EI spectra obtained with the moving-belt interface. These 14 alkaloids represented an estimated 95% (by weight) of the total alkaloids present. Raynor et al. described a further application of this interface to natural product identification in their characterization of phytoecdysteroids from *Silene nutans* and *Silene otites* [72]. Though ionization was initiated by high voltage discharge rather than high energy electrons, the methanol mobile-phase modifier again provided methanol CI.

As discussed earlier in the DFI approach to SFC/MS, double-focussing sector mass analyzers provide advantages over quadrupole instruments, such as the ability to perform high resolution mass analysis. Chapman interfaced packed SFC to a double-focussing sector instrument using a modified thermospray interface [70]. The modification allowed a fused silica flow restrictor to pass through the heated stainless steel "vaporizer" directly into the thermospray ion source. It also allowed the coaxial addition of solvents to the thermospray source. As in the work of Games and coworkers, Chapman found that the mobile phase, and especially coaxially added solvents, influenced ionization within the thermospray interface. For example, methylene chloride dramatically enhanced the response for carotinoid hydrocarbons, though it did not otherwise affect the spectra.

Cousin and Arpino [27] and Smith and Udseth [25] described high-flow-rate SFC/MS interfaces which resembled the thermospray interface in the sense that the entire effluent was introduced directly to the CI source to which a mechanical pump was attached. Smith and Udseth abandoned this approach due to poor response in favor of the high-flow-rate interface described earlier [25]. Cousin and Arpino pursued the characterization of their interface, using polynuclear aromatic hydrocarbons, low molecular weight triglycerides, and fatty acid methyl esters as probes. They found that a mixture of NH_3 CI and CO_2 CE was helpful in providing both molecular weight and structural information. They did observe that S/N and mass range were reduced at high (100 μL/min) mobile phase flow rates. The authors suggest that performance would be improved had high-energy-electron (> 100 eV) or discharge ionization been available, as they are in traditional thermospray interfaces. Arpino and Cousin used the interface to study ion-molecule reactions of nonretained solutes (coeluting with the injection solvent) with ions formed from the injection solvent [73].

9.5 Supercritical Fluid Injection/Mass Spectrometry

The introduction of supercritical fluid solutions into the ion source of a mass spectrometer without a prior chromatographic separation is referred to here as supercritical fluid injection/mass spectrometry (SFI/MS). The use of SFI/MS will certainly grow as does laboratory-scale supercritical fluid extraction. In favorable cases, the selectivity of the mass spectrometer alone is sufficient for identification and even quantitation of analytes in a supercritical fluid extract.

An excellent example of the use of mass spectrometric selectivity in SFI/MS was provided by Kalinoski et al. [41]. They spiked wheat samples with trichothecene mycotoxins and extracted the toxins with a supercritical mixture of CO_2 and 5% (v/v) 2-propanol. The extract was injected directly into the CI source of a double-focussing sector instrument using their high-voltage, direct-fluid-introduction interface probe. The authors chose to introduce ammonia to the ion source in order to obtain ammonia CI, rather than use the 2-propanol modifier as the source of CI reagent. Figure 12

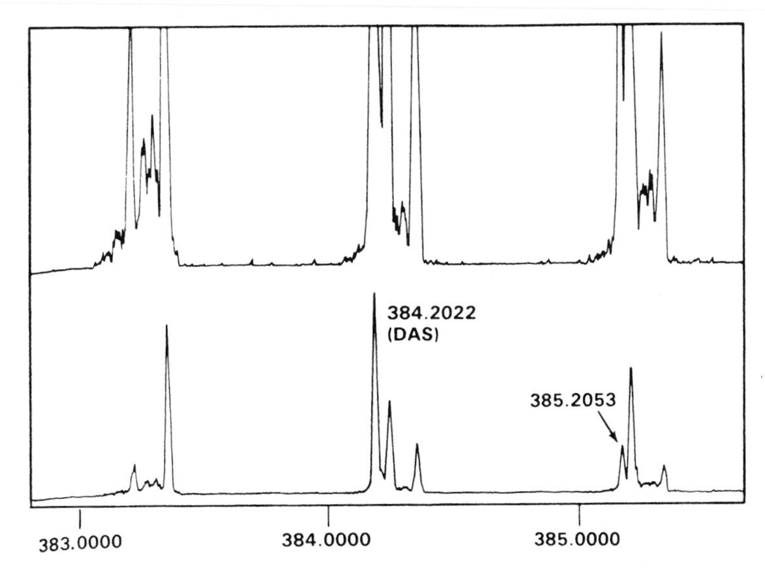

Fig. 12. An oscillographic trace of the high-resolution mass spectrum of a supercritical CO_2/2-propanol (95:5) extract of wheat containing diacetoxyscirpenol (DAS) and T-2 toxin. SFI/ammonia CIMS was used at a mass resolving power of 7400 (5% valley) on a VG ZAB mass spectrometer. Reproduced from Ref. [41] with permission from Elsevier Scientific Publishing Company

shows the ammonium-adduct-ion region for diacetoxyscirpenol (DAS, MW 366). High resolution mass analysis was successful in determining DAS in the presence of other species of the same nominal mass. In a previous publication, Kalinoski et al. used SFI and the selectivity of tandem mass spectrometry with a triple quadrupole instrument to unambiguously identify mycotoxins in supercritical fluid extracts [14]. Wright and Smith also studied the mass spectral behavior of porphyrins by SFI/MS of their solutions in supercritical ammonia [68].

Pinkston et al. suggested the use of SFI/MS of solutions of polysiloxanes for calibration and tuning [74]. Figure 13 shows the NH_3 CI SFI/MS spectrum of a solution of poly(methyl-3,3,3-trifluoropropylsiloxane) in CO_2 modified with 5% 2-propanol. The mixture provided a simple spectrum covering a mass range (m/z 1500–4000) difficult to reach using traditional calibration compounds desorbed from a direct-insertion probe, but appropriate for SFC/MS using modern "high mass" quadrupole instruments.

Lubman and coworkers used the selectivity of resonant two-photon ionization (R2PI) to probe analytes in SFI/supersonic beam mass spectrometry [75, 76]. Species of interest were dissolved in supercritical CO_2 as it flowed over the solid sample. The solution was injected through a pulsed valve into the laser ionization region [76]. The resulting ions were mass analyzed using a simple time-of-flight mass spectrometer. The rotational-vibrational cooling of the analyte molecules provided by carbon dioxide

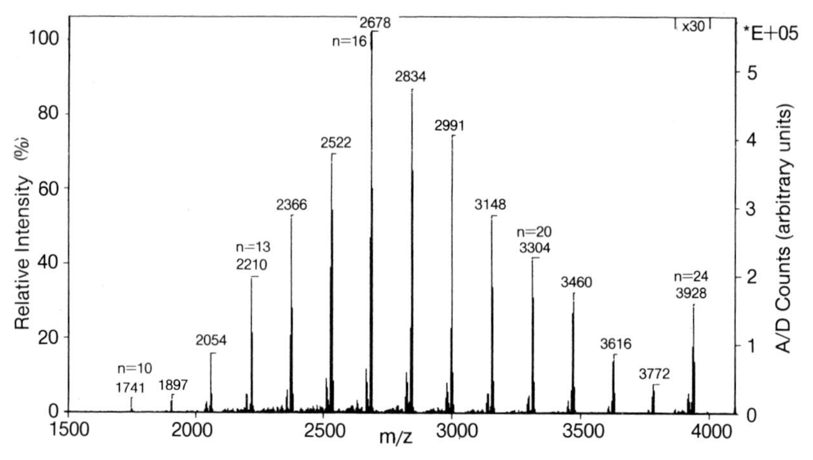

Fig. 13. Ammonia CI SFI/MS spectrum of 3% poly(methyl-3,3,3-trifluoropropylsiloxane) in CO_2 modified with 5% 2-propanol. Reproduced from Ref. [74] with permission from the American Chemical Society

was not as effective as that provided by argon. However, the spectral features obtained with CO_2 expansions were still sharp enough to enable identification of analytes.

The full potential of SFE/SFI/MS/MS is yet to be realized. As the advantages of analytical scale SFE over traditional extraction methods become known, and as tandem mass spectrometers become more widespread, this powerful combination will undoubtedly grow in use. It should be especially useful for rapid, targeted compound analysis.

9.6 Conclusion

Capillary and packed SFC/MS each present unique advantages for certain types of applications. A distinct advantage of capillary SFC/MS is that capillary SFC/FID methods (mobile phase, flow rate, and column) can be directly transferred to direct-fluid-injection capillary SFC/MS instruments. Both EI-type and all types of CI can be obtained when unmodified CO_2 is the mobile phase. Packed SFC/MS has an advantage in speed of analysis and less stringent dead-volume and injection-volume constraints. These advantages make the two techniques complementary. For example, capillary SFC/MS may be best suited for the identification of components of complex mixtures, while packed SFC/MS may be best suited for targeted compound analysis.

The development of most new analytical methods seem to proceed through a series of stages characterized initially by excitement and optimistic speculation about the promise of the technique, followed by disillusionment, followed by a period of gradual progress and application [77]. SFC is moving into the third stage, with SFC/MS close behind. There are now areas, such as the characterization of nonionic surfactants, where SFC exhibits clear advantages over GC and HPLC. The rate of progress has certainly been slower than instrument manufacturers hoped, but the number of SFC/MS practitioners will grow as the use of SFC and analytical-scale SFE becomes more widespread.

Acknowledgements. I would like to thank T.L. Chester and R.P. Oertel for help in reviewing the manuscript and for useful discussions.

References

1. Novotny M, Springston SR, Peaden PA, Fjeldsted JC, Lee ML (1981) Anal Chem 53:407A
2. Springston SR, Novotny M (1981) Chromatographia 14:679
3. Peaden PA, Fjeldsted JC, Lee ML, Springston SR, Novotny M (1982) Anal Chem 54:1090
4. Smith RD, Fjeldsted JC, Lee ML (1982) J Chromatogr 247:231
5. Smith RD, Felix WD, Fjeldsted JC, Lee ML (1982) Anal Chem 54:1883
6. Milne TA (1969) Int J Mass Spectrom Ion Phys 3:153
7. Giddings JC, Myers MN, Wahrhaftig AL (1970) Int J Mass Spectrom Ion Phys 4:9
8. Randall LG, Wahrhaftig AL (1978) Anal Chem 50:1703
9. Randall LG, Wahrhaftig AL (1981) Rev Sci Instrum 52:1283
10. Gouw TH, Jentoft RE, Gallegos EJ (1979) J High Pressure Sci Technol 6:583
11. Smith RD, Udseth HR (1983) Anal Chem 55:2266
12. Smith RD, Fjeldsted J, Lee ML (1983) Int J Mass Spectrom Ion Phys 46:217
13. Smith RD, Kalinoski HT, Udseth HR, Wright BW (1984) Anal Chem 56:2476
14. Kalinoski HT, Udseth HR, Wright BW, Smith RD (1986) Anal Chem 58:2421
15. Huang EC, Jackson BJ, Markides KE, Lee ML (1988) Anal Chem 60:2715
16. Huang EC, Jackson BJ, Markides KE, Lee ML (1988) Chromatographia 25:51
17. Hawthorne SB, Miller DJ (1987) J Chromatogr 388:397
18. Zaugg SD, Deluca SJ, Holzer GU, Voorhees KJ (1987) HRC CC J High Resolut Chromatogr Chromatogr Comm 10:100
19. Hawthorne SB, Miller DJ (1988) Fresenius' Z Anal Chem 330:235
20. Lee ED, Henion JD (1986) HRC CC J High Resolut Chromatogr Chromatogr Comm 9:172
21. Owens GD, Burkes LJ, Pinkston JD, Keough T, Simms JR, Lacey MP (1988) ACS Symp Ser 366:191
22. Pinkston JD, Owens GD, Burkes LJ, Delaney TE, Millington DS, Maltby DA (1988) Anal Chem 60:962
23. Pinkston JD, Bowling DJ, Delaney TE (1989) J Chromatogr 474:97
24. Reinhold VN, Sheeley DM, Kuei J, Her G (1988) Anal Chem 60:2719
25. Smith RD, Udseth HR (1987) Anal Chem 59:13
26. Kalinoski HT, Hargiss LO (1989) Presented at the 6th (Montreux) Symposium on LC/MS (LC/MS; SFC/MS; CZE/MS; MS/MS), July 19–21, 1989, Ithaca, New York, USA
27. Cousin J, Arpino PJ (1987) J Chromatogr 398:125
28. Sheeley DM, Reinhold VN (1989) Presented at the 37th ASMS Conference on Mass Spectrometry and Allied Topics, May 21–26, 1989, Miami Beach, Florida, USA

29. Blum W, Grolimund K, Jordi PE, Ramstein P (1988) HRC&CC J High Resolut Chromatogr Chromatogr Comm 11:441
30. Roach JAG, Sphon JA, Easterling JA, Calvey EM (1989) Biomed Environ Mass Spectrom 18:64
31. Smith RD, Udseth HR, Kalinoski HT (1984) Anal Chem 56:2971
32. Holzer G, Deluca S, Voorhees KJ (1985) HRC CC J High Resolut Chromatogr Chromatogr Comm 8:528
33. Lee ED, Hsu S, Henion JD (1988) Anal Chem 60:1990
34. Chester TL, Pinkston JD, Innis DP, Bowling DJ (1989) J Microcol Sep 1:182
35. Pinkston JD, Delaney TE, Bowling DJ (1990) J Microcol Sep 2:181
36. Wenclawiak B, Sichtermann W, Benninghoven A (1989) Fresenius' Z Anal Chem 335:549
37. Frew NM, Johnson CG, Bromund RH (1988) ACS Symp Ser 366:208
38. Kalinoski HT, Udseth HR, Wright BW, Smith RD (1987) J Chromatogr 400:307
39. Berry AJ, Games DE, Mylchreest IC, Perkins JR, Pleasance S (1988) HRC CC J High Resolut Chromatogr Chromatogr Comm 11:61
40. Kalinoski HT, Hargiss LO (1989) J Chromatogr 474:69
41. Kalinoski HT, Udseth HR, Chess EK, Smith RD (1987) J Chromatogr 394:3
42. Wilkins CL, Chowdhury A, Nuwaysir LM, Coates ML (1989) Mass Spectrom Rev 8:67
43. Freiser BS (1988) Tech Chem (NY) 20:61
44. Chiarelli MP, Gross ML (1988) Anal Appl Spectrosc 263
45. Lee ED, Henion JD, Cody RB, Kinsinger JA (1987) Anal Chem 59:1309
46. Laude DA Jr, Pentoney SL, Griffiths PR, Wilkins CL (1987) Anal Chem 59:2283
47. Stafford GC Jr, Kelley PE, Syka JEP, Reynolds WE, Todd JFJ (1984) Int J Mass Spectrom Ion Proc 60:85
48. Hemberger PH, Moss JD, Kaiser RE, Louris JN, Amy JW, Cooks RG, Syka JEP, Stafford GC (1989) Presented at the 37th ASMS Conference on Mass Spectrometry and Allied Topics, May 21–26, 1989, Miami Beach, Florida, USA
49. Todd JFJ, Mylchreest IC, Berry AJ, Games DE, Smith RD (1988) Rapid Comm in Mass Spectrom 2:55
50. McFadden WH, Schwartz HL, Evans S (1976) J Chromatogr 122:389
51. Berry AJ, Games DE, Perkins JR (1986) Anal Proc (London) 23:451
52. Berry AJ, Games DE, Perkins JR (1986) J Chromatogr 363:147
53. Games DE, Berry AJ, Mylchreest IC, Perkins JR, Pleasance S (1987) Anal Proc (London) 24:371
54. Balsevich J, Hogge LR, Berry AJ, Games DE, Mylchreest IC (1988) J Nat Prod 51:1173
55. Ramsey ED, Perkins JR, Games DE, Startin JR (1989) J Chromatogr 464:353
56. Niessen WMA, de Kraa MAG, Verheij ER, Bergers PJM, La Vos GF, Tjaden UR, van der Greef J (1989) Rapid Comm in Mass Spectrom 3:1
57. Henion JD, Edlund PO, Lee ED, McLaughlin L (1988) Presented at the 36th ASMS Conference on Mass Spectrometry and Allied Topics, June 5–10, 1988, San Francisco, CA, USA
58. Edlund PO, Henion JD (1989) J Chromatogr Sci 27:274
59. Keough T, Pinkston JD (1990) unpublished results
60. Wenclawiak B (1990) Fresenius J Anal Chem 337:129
61. Crowther JB, Henion JD (1985) Anal Chem 57:2711
62. Matsumoto K, Tsuge S, Hirata Y (1986) Anal Sci 2:3
63. Matsumoto K, Tsuge S, Hirata Y (1986) Chromatographia 21:617
64. Matsumoto K, Tsuge S, Hirata Y (1987) Shitsuryo Bunseki 35:15
65. Kalinoski HT, Smith RD (1988) Anal Chem 60:529
66. Wright BW, Udseth HR, Chess EK, Smith RD (1988) J Chromatogr Sci 26:228
67. Kalinoski HT, Wright BW, Smith RD (1988) Biomed Environ Mass Spectrom 15:239
68. Wright BW, Smith RD (1989) Org Geochem 14:227
69. Vestal ML (1983) Mass Spectrom Rev 2:447
70. Chapman JR (1988) Rapid Comm in Mass Spectrom 2:6
71. Berry AJ, Games DE, Mylchreest IC, Perkins JR, Pleasance S (1988) Biomed Environ Mass Spectrom 15:105

72. Raynor MW, Kithinji JP, Bartle KD, Games DE, Mylchreest IC, Lafont R, Morgan E, Wilson ID (1989) J Chromatogr 467:292
73. Arpino PJ, Cousin J (1987) Rapid Comm in Mass Spectrom 1:29
74. Pinkston JD, Owens GD, Petit EJ (1989) Anal Chem 61:775
75. Sin C, Pang H, Lubman DM, Zorn J (1986) Anal Chem 58:487
76. Pang H, Sin C, Lubman DM, Zorn J (1986) Anal Chem 58:1581
77. Laitinen HA (1973) Anal Chem 45:1205

10 Supercritical Fluid Chromatography with FT-IR Detection

LARRY T. TAYLOR and ELIZABETH M. CALVEY

10.1 Introduction

Both conventional HPLC and GC detectors have proven to be compatible with Supercritical Fluid Chromatography (SFC). For example, flame ionization, electron capture, flame photometric, thermionic, ultraviolet and fluorescence detection [1, 2] have all been successfully demonstrated. Various efforts have been made to couple spectrometric detectors with chromatographic systems in order to gain more specific information regarding eluting components. Fourier Transform infrared (FT-IR) [3], atomic emission, nuclear magnetic resonance and mass [4] spectrometry have been interfaced to SFC with varying degrees of success during the 1980s. The focus of this chapter is SFC/FT-IR. The FT-IR detector is constrained by two major problems: mid-IR absorption by most chromatographically compatible mobile phases and relatively low FT-IR sensitivity compared to some other more established detectors. In order to minimize these problems, various ingenious interface designs have been explored. These designs appear to vary greatly, but they can be classified into two approaches: solvent elimination coupled with transmission or reflectance IR, and flow cell coupled with transmission or attenuated total reflectance IR. Each approach has a unique set of characteristics that makes it attractive for certain applications.

10.2 Flow Cell Approach

Carbon dioxide is a viable mobile phase for SFC/FT-IR due to its infrared transparency. Only those regions from 3475 to 3850 and from 2040 to 2575 cm^{-1} and below 800 cm^{-1} are completely lost because of strong absorption by CO_2. Another area of the spectrum where information is potentially lost or reduced is between 1200 and 1400 cm^{-1}, where increased absorption by CO_2 is caused by Fermi resonance whose magnitude is a function of CO_2 density. The increase in absorptivity of this region that occurs as density increases causes severe baseline drift which may mask solute peaks. Because of this phenomenon, many of the initial flow cell studies with on-line FT-IR involved isobaric conditions and essentially provided increased support for the solvent elimination approach, which could

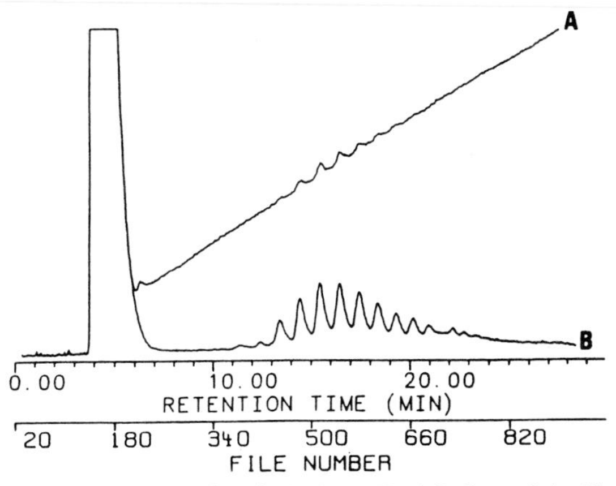

Fig. 1. Compensation for carbon dioxide density gradient in SFC/FT-IR. (*A*) Gram-Schmidt reconstructed chromatogram using 10 basis vectors from start of run; (*B*) same data with an additional basis vector taken from file 900 (29.12 min) added to the basis set. Reprinted with permission from Ref. [6]. Copyright American Chemical Society

easily accommodate a variety of density programs. However, certain mobile phases, which do not exhibit any infrared absorbance bands such as supercritical xenon, were viable for flow cell interfaces with density programming [5].

Wieboldt and Hanna [6] have overcome the undesirable baseline rise, due to increased CO_2 absorptivity as a function of a density increase, in supercritical fluid chromatograms by using Gram-Schmidt orthogonalization with an augmented basis vector set. As shown in Fig. 1, the addition of a vector from the high density region of the paraffin wax mixture chromatogram deconvolutes the chromatographic peaks from the baseline drift caused by the density program and enhances detection of the chromatographic peaks.

Demonstration of this method of data treatment can be seen by the separation of a methylene chloride mixture of four pesticides (e.g. Aldicarb, methomyl, captan, and phenmedipham) on a poly(methylsiloxane) open tubular capillary column (10 m × 100 μm) with density programming at 100 °C [7]. FT-IR spectra were recorded at 8 cm^{-1} resolution with 8 scans co-added per file. Figure 2 shows the chromatogram, generated from approximately 50 ng of each component injected, that was reconstructed from the total IR response. Figure 3 is the IR spectrum of the component eluting in the first peak, Aldicarb, obtained by co-adding 96 scans. Several chemical features are immediately apparent from the spectrum. The strong band at 1762 cm^{-1} is caused by the carbonyl C=O stretch. The presence of the C–O stretching band at 1217 cm^{-1} indicates an ester functionality. The band at 3460 cm^{-1} is definitive evidence for a secondary N–H stretch. The additional band at 1507 cm^{-1} indicates that the nitrogen is part of an amide group. The two blank portions of the spectrum are the regions in

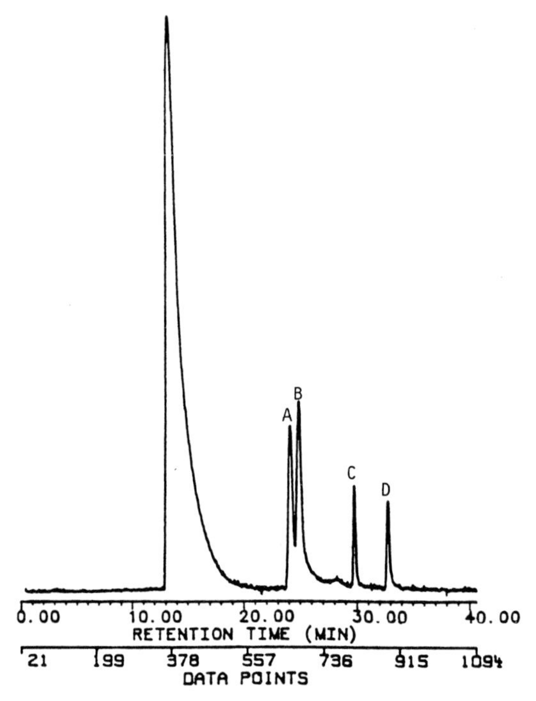

Fig. 2. Separation of a carbamate pesticide mixture by SFC/FT-IR. Mobile phase: supercritical CO_2; linear velocity, ~1.4 cm/s; density program: 6.0 min hold at 0.180 g/ml, then to 0.360 g/ml at 0.010 g/ml/min, then to 0.600 g/ml at 0.040 g/ml/min, followed by 10.0 min hold; injection: 200 nl; split ratio: 22:1; column: 10 m × 100 μm SB-Methyl-100 capillary column; Oven temperature: 100 °C. Peaks: A = aldicarb, B = methomyl, C = captan, D = phenmedipham. Reprinted with permission from Ref. [7]

which the supercritical CO_2 mobile phase absorbs all the available IR energy.

The cell design employed in the above study is also applicable to packed column SFC and in terms of chromatographic performance should perform better because peak volumes and cell volumes are more compatible. The dimensions of the commercially available (Nicolet) flow cell are 0.60 mm i.d. × 5 mm pathlength which provides a cell volume of 1.4 μl. The transfer lines from the chromatographic column and to the restrictor are made from fused silica (0.5 m × 50 μm i.d.). The flow cell design as reported by Wieboldt [8] was a compromise between the conflicting requirements of an absorbance detector (longer pathlength) and a chromatographic detector (small cell volume). Specifically, the cell volume was five times greater than the theoretically allowable detector cell volume for a 20 m × 100 μm i.d. capillary column, with a plate height of 0.6 d_c (internal column diameter) and a k′ value of 1. The optics of the detector system dictated the cell diameter, therefore, any changes in the cell volume could only be achieved at the expense of detector pathlength and sensitivity (i.e. shorter pathlength,

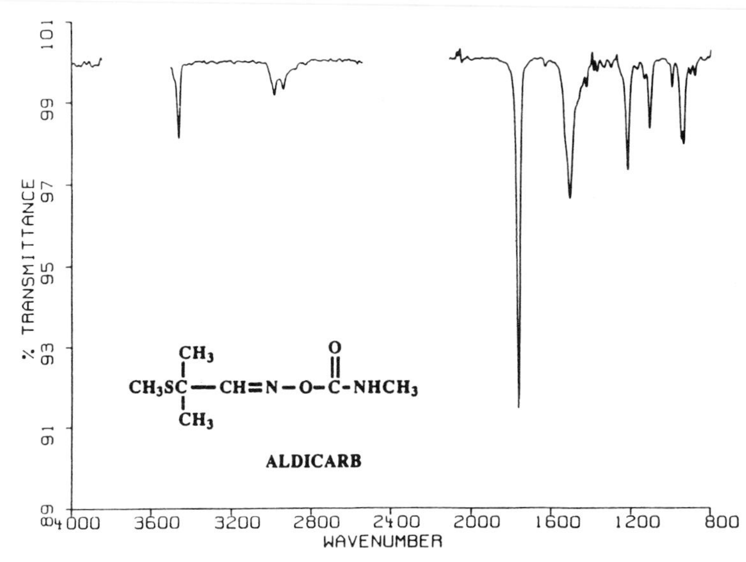

Fig. 3. On-line SFC/FT-IR spectrum of aldicarb (peak A in Fig. 5). Conditions: 8 cm^{-1} resolution; 8 scans co-added per file; 12 files co-added. Reprinted with permission from Ref. [7]

less sensitivity; longer pathlength, less throughput). Due to the increased absorption of the Fermi bands in CO_2, a 5-mm pathlength was found to be the maximum practical length when working at high densities. The transfer line and the flow cell were independently thermostated at 31 °C because an improvement in peak shape was expected when the flow cell was at sub-critical conditions due to peak compression as the density of the carrier fluid will increase within the cell.

This flow cell SFC/FT-IR interface has been employed with a variety of mixtures. Five steroids were separated [9] using a cyanopropyl polysiloxane capillary column. Sequential detection via flame ionization after passage through the FT-IR flow cell yielded similar chromatographic traces for this mixture (Fig. 4). Peracetylated nitrogen derivatives of seven monosaccharides have been analyzed [10, 11] using both 1.0 mm, i.d. packed and 100 μm, i.d. open tubular columns. Multiple derivatives for each monosaccharide were distinguishable by the various carbonyl stretching vibrational modes observed in the on-line IR spectra. Prior to this study gas chromatography coupled with flame ionization detection of the derivative mixture had suggested a single product per monosaccharide. Mass spectrometric data (SFC/MS) were needed to confirm the presence of the nitrile and oxime moieties because the nitrile stretch was masked by CO_2 absorbance and the C=N stretch (oxime) was took weak to be observed at the levels analyzed. The utility of this flow cell interface has been further shown by the analysis of pyrethrins [12], ureas, benzamides, sulfonamides [13], propellant components [14] and tobacco extracts [15].

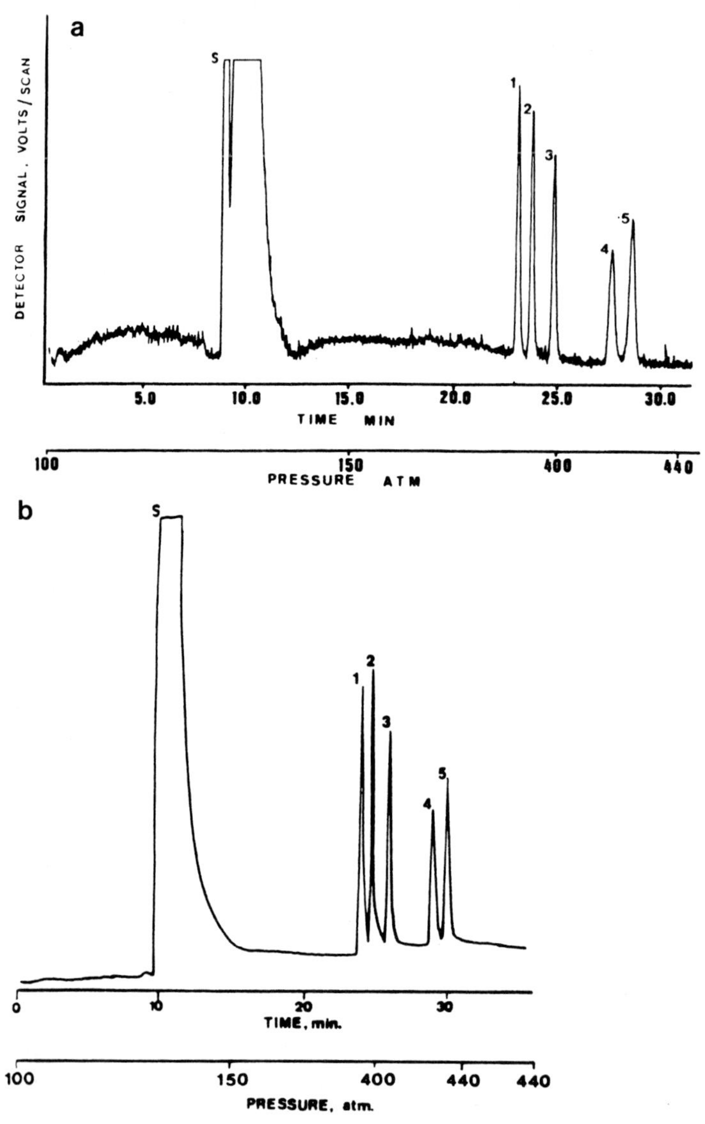

Fig. 4a, b. Separation of model steroid mixture (**a**) by SFC/FT-IR and (**b**) by SFC/FID (post FT-IR). Separation performed on SB-cyanopropyl-25 column (10 m × 100 μm, i.d.) at 60 °C with 100% CO_2. *S* = CH_2Cl_2, *1* = progesterone, *2* = testosterone, *3* = 17-hydroxy-progesterone, *4* = 11-deoxycortisol, *5* = corticosterone. Reprinted with permission from Ref. [8]. Copyright Vieweg Verlagsgesellschaft

The addition of polar modifiers to increase the solvent strength of CO_2 or to deactivate the stationary phase reduces the applicability of on-line FT-IR for obtaining identifiable spectra. Jordan and Taylor [16] showed that with a 5-mm pathlength cell, the addition of as little as 0.2% methanol reduced the accessible IR windows to 3400–2900 cm^{-1}, 2800–2600 cm^{-1}, 2100–1500 cm^{-1}, and 1200–1100 cm^{-1}. Morin and co-workers [17], using a 10-mm pathlength cell with an 8-μl volume, studied the IR transparency of CO_2 with the addition of various polar modifiers under subcritical conditions. While the addition of polar modifiers caused a severe loss of available IR windows, specific frequencies could still be selectively monitored. For example, the carbonyl and carbon-carbon double bond stretching regions always remained transparent with methanol and acetonitrile as modifiers. The use of CD_3CN as a modifier permitted monitoring of the C–H stretching region (2900–3100 cm^{-1}), and with less than 9% CD_3CN added the aliphatic CH_2 and CH_3 bending region (1600–1400 cm^{-1}) could also be monitored.

Other considerations when deciding on the applicability of a flow cell SFC/FT-IR interface are the minimum identification limit (MIL) defined as [18] the quantity of compound required for identification by spectral interpretation or computer search and the injected minimum detectable quantity (IMDQ) defined as [19] the quantity of material which must be injected onto the column of choice to yield an infrared response three times the noise level. In contrast to solvent elimination methods, in which the number of scans can be increased to reduce spectral noise, flow cell methods provide a finite number of scans per peak since spectral acquisition is performed in real time. Maximum sensitivity will not be realized if data are taken only at the peak maximum (or at a fixed time), since peaks with higher k′ values will be broader, and a smaller fraction of the total analyte will be sampled. Consequently, in order to achieve maximum sensitivity for all analytes, a method to optimize the signal-to-noise ratio (S/N) of the infrared spectrum generated in the flowing experiment must be adopted. Both from a theoretical treatment [19] and from experimental data [20] it has been shown that maximum S/N is realized when ±1.37 standard deviations of the chromatographic peak (~75%) are sampled. Certain spectroscopic and molecular parameters also affect detectability. For example, if the molar absorptivity of the vibrational mode is very large, detection limits will be significantly lower, provided the noise level is invariable.

Shah and co-workers [9] have indicated that for a strongly absorbing compound such as caffeine, the IMDQ is approximately 2 ng. Caffeine was eluted with 100% CO_2 from a 25% cyanopropyl polysiloxane column in approximately 17.5 min with a k′ of 0.97. Various quantities of caffeine (250, 50, 25, 5 and 2.5 ng) were injected and the absorbance of the intense carbonyl peak (from the on-line FT-IR spectrum) was correlated with the amount injected. For each injection 12 files (48 scans) were co-added across the caffeine chromatographic peak to acquire the IR spectrum of greatest S/N. An S/N (peak to peak ±50 cm^{-1} from the reference peak) of greater than three was achieved for as little as 2.5 ng injected. Figure 5 illustrates a portion of the on-line FT-IR spectrum generated under these conditions. With spectral detectors such as FT-IR, identification limits may be more useful than detection limits. Wieboldt et al. [8] determined that the MIL

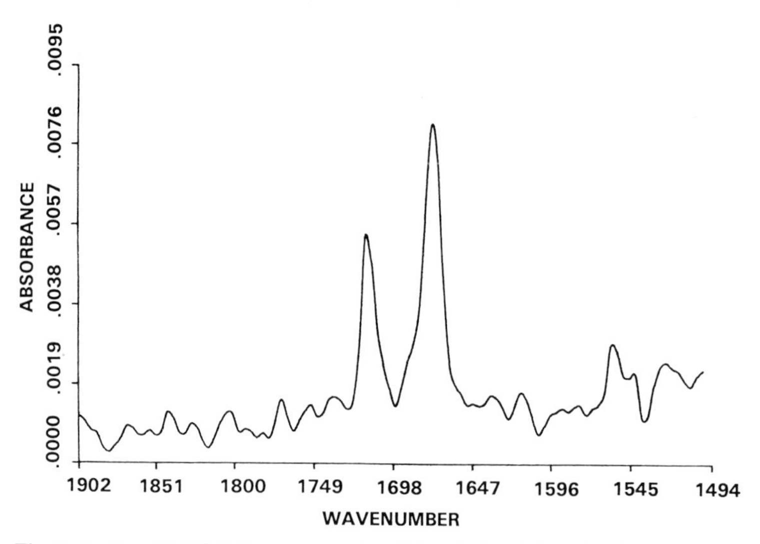

Fig. 5. On-line SFC/FT-IR spectrum of caffeine (2.5 ng injected). 12 co-added files, 4 scans/file, 1 file/s. SFC conditions: SB-cyanopropyl-25 column (10 m × 100 μm, i.d.) at 60 °C with 100% CO_2; linear pressure programming (100–175 atm/15 min, 175–400 atm/5 min). Reprinted with permission from Ref. [8]. Copyright Vieweg Verlagsgesellschaft

for methyl palmitate was 10 ng on column (10 m × 100 μm, 0.50 μm film polydimethylsiloxane).

10.3 Solvent Elimination Approach

In the solvent elimination method, each chromatographic peak is deposited from the end of the restrictor (connected to the end of the column by a heated transfer line) onto a small area of an IR-transparent support (Fig. 6). The mobile phase is then allowed to evaporate away. The support is then positioned in the IR beam and spectra are collected, with as many scans as are necessary to obtain adequate S/N. Stepping or continued movement of the support beneath the restriction outlet ensures that chromatographed peaks are separated spatially.

Cooling of the support may be necessary if fairly volatile constituents are to be collected. In many cases this avoids the deposition of compounds aspirated as liquids which would otherwise spread over a larger area of window surface [21]. In this regard, capillary SFC with matrix isolation FT-IR has been demonstrated [22] using CO_2 with 0.5 mole % CCl_4 as a mobile phase. The eluting analytes were isolated in a CCl_4 matrix on a low temperature gold disk held at 150 K, the lowest temperature at which CO_2 was not deposited on the disk. IR spectra were obtained above 900 cm^{-1}. The

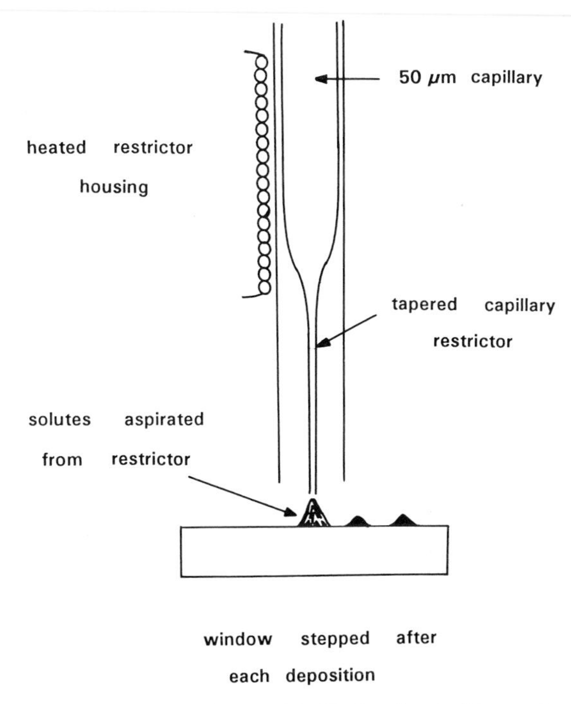

Fig. 6. Basic design of a solvent elimination interface for SFC/FT-IR. Reprinted with permission from Ref. [27]. Copyright Wiley

MI FT-IR spectrum of approximately 81 ng of 1,4-naphthoquinone obtained by co-adding 16 scans was reported. A peak at 2337 cm^{-1} was observed to be due to CO_2 trapped in the carbon tetrachloride matrix. Although all six components in a test mixture were chromatographed as evidenced by the FID response only two components, *n*-octanol and *N,N*-dimethylaniline, were isolated on the disk in sufficient amounts to obtain usable spectra.

The type of IR method used for detection depends on the substrate employed. Fuoco et al. [23] compared a variety of sampling techniques for combined SFC/FT-IR with mobile phase elimination. The techniques compared were: 1) Conventional transmission using a ZnSe plate; 2) External reflection spectrometry using a ZnSe plate; 3) Reflection-absorption spectrometry using a flat Al substrate; 4) Diffuse reflection spectrometry using a fine NaCl powder on a flat metallic surface (Al); 5) Diffuse reflection spectrometry using a fine NaCl powder on an IR-transparent surface (ZnSe); and 6) Diffuse transmittance spectrometry using NaCl powder on a smooth ZnSe surface. The comparisons were made by measuring the spectrum of acenaphthenequinone on a stationary rather than a moving substrate. The data showed that conventional transmission spectrometry appears to be the technique of choice. Fewer spectral artifacts are obtained in conventional transmission spectra than in reflectance-absorption spectra. The band intensities of the diffuse transmission spectrum were more intense than those in the conventional transmission spectrum. But, the SNR of the conventional transmission spectrum was superior to that of diffuse transmission spectrum. The relative band intensities of the conventional transmission spectra showed a dependence on amount of sample deposited especially at small deposited amounts. This change was postulated to be caused by the loss of transmitted radiation by external reflection at the wavelength of the absorption bands. These workers note that this is the first reported SFC/FT-IR spectra of a subnanogram (800 pg) injected quantity of an SFC elute.

Initial studies involving the solvent elimination approach concentrated on the use of packed columns. Fujimoto and co-workers [24] used a continuously moving potassium bromide crystal and obtained IR spectra by using beam condenser optics. Shafer and co-workers [25] deposited the effluent from a microbore packed column onto a potassium chloride powder strip and performed diffuse reflectance IR spectrometry (DRIFT). In order to extend the method to a capillary column, Pentoney and co-workers [21] have recently used a microscope accessory to match the IR beam size to the area occupied by the column effluent. This feature was found to be desirable since the sensitivity of the solvent elimination method is maximized by depositing the column effluent over as small a spot area as possible. Raynor and co-workers [26, 27] used a capillary column (100 μm, i.d.) and obtained spot sizes less than 300 μm in diameter with the end of the restrictor approximately 50 μm above the surface of a potassium bromide

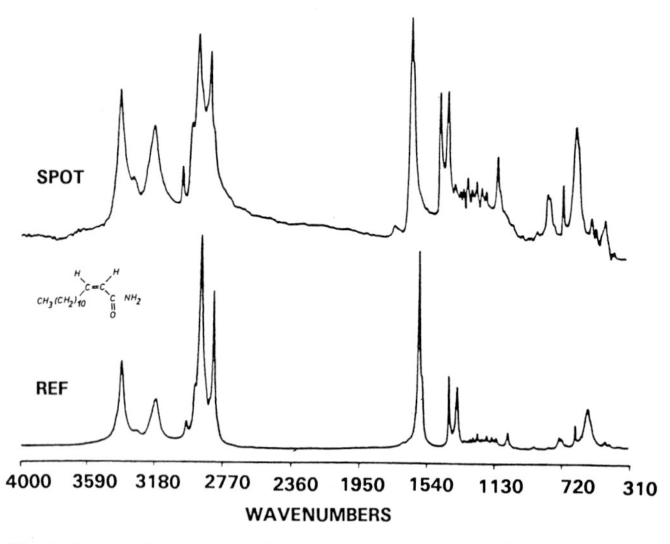

Fig. 7. Spot and reference infrared spectra of erucamide. Reference material was ground into a potassium bromide disk. Spectra were collected after 1000 co-added scans at a resolution of 4 cm^{-1}. Reprinted with permission from Ref. [27]. Copyright American Chemical Society

window. Identifiable spectra for several separated polymer additives with the deposition of 100 ng of each component were obtained. Figure 7 indicates the spot and reference spectra of erucamide (i.e., a long chain mono-olefin primary amide). The spot spectra represented 1000 scans measured with a resolution of 4 cm^{-1} and accumulated in 4 min. For the two most retained compounds in this study (Irqanox 3114 and 1010) the aliphatic C−H stretching absorption in the 3100−2800 cm^{-1} region was stronger than that obtained for reference spectra. This increase in absorption was explained by the presence of hydrocarbon impurities in the CO_2.

Flame ionization is normally used as a secondary detector to ascertain when the substrate should be moved in order to accommodate the elution of another component. The use of this detector requires that the column effluent be split between the FT-IR and flame ionization detectors. Raynor [27] has indicated that this splitting could cause a compromise in chromatography because in some cases column overload was necessary in order to obtain enough analyte for acquisition of IR data. He suggested that a UV cell in-line could reduce the quantity of solute injected onto the column by eliminating the splitting requirement. If a UV detector is used, however, compounds without chromophores could be codeposited unless a continuously moving crystal was used to collect the column effluent.

As mentioned previously, the solvent elimination technique provides a means to investigate other less infrared compatible supercritical fluids as potential mobile phases. The infrared transparency of CF_2Cl_2 is poor,

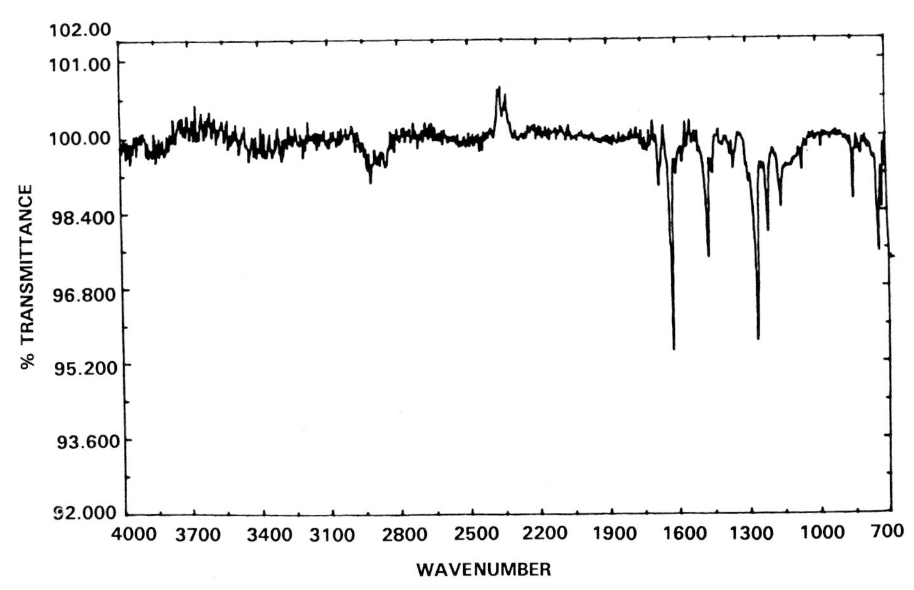

Fig. 8. Transmittance spectrum of 1,8-dihydroxy-anthraquinone after elution from SFC capillary column using CF_2Cl_2 as the mobile phase and deposited onto a ZnSe window. Reprinted with permission from Ref. [28]

therefore, not a viable mobile phase using the flow cell technique. Figure 8 shows the transmittance spectrum of 1,8-dihydroxyanthraquinone after elution from a capillary column using CF_2Cl_2 as the solvent [28]. By eliminating the solvent, absorbance bands in the fingerprint region can be easily observed, although impurities in the mobile phase may interfere.

10.4 Summary

Since the introduction of the first commercially available SFC system in the early 1980s, the development of the hyphenated SFC/FT-IR technique has attained a high level of sophistication. The many types of interfaces that are being actively investigated for SFC/FT-IR are based on those developed for HPLC. The advantages for the solvent elimination approach include: 1) accommodates a variety of mobile phases including CO_2 modified with polar solvents such as methanol; 2) permits the use of standard condensed phase reference spectra; and 3) number of scans can be increased to reduce spectral noise and increase sensitivity. The advantages for the flow cell approach include: 1) the entire effluent stream is monitored so all sample components are detected intact; 2) other detectors such as the mass spectrometer may be placed in series after the FT-IR; and 3) the interface is mechanically simple. The disadvantages of the solvent elimination approach include: 1) interface designs tend to be mechanically complex; 2) removal of the mobile phase in some instances can drive off volatile components; and 3) mobile phase impurities may be co-deposited onto the substrate and cause interferences in the collected spectra. The disadvantages of the flow cell approach include: 1) separate spectral libraries may be required for various mobile phases since complete spectra are not feasible; 2) usable mobile phases are limited when desiring identifiable spectra; and 3) lower sensitivity can be expected in most cases.

Acknowledgment. E. M. Calvey gratefully acknowledges a long-term training appointment from the US Department of Health and Human Services. The financial assistance of the US Environmental Protection Agency is deeply appreciated.

References

1. Bornhop DJ, Wangsgaard JG (1989) J Chromatogr Sci 27:293
2. Richter BE, Bornhop DJ, Swanson JT, Wangsgaard JG, Anderson MR (1989) J Chromatogr Sci 27:303
3. Taylor LT, Calvey EM (1989) Chem Rev 89:321
4. Smith RD, Udseth HR (1987) Anal Chem 59:13

5. French SB, Novotny M (1986) Anal Chem 58:164
6. Wieboldt RC, Hanna DA (1987) Anal Chem 59:1255
7. Wieboldt RC (1987) Nicolet FT-IR Application Note AN-8705, March 1987
8. Wieboldt RC, Adams GE, Later DW (1988) Anal Chem 60:2422
9. Shah S, Ashraf-Khorassani M, Taylor LT (1988) Chromatographia 25:631
10. Calvey EM, Taylor LT, Palmer JK (1988) J High Resolut Chromatogr Chromatogr Comm 1:294
11. Calvey EM, Roach JAG, Taylor LT, Palmer JK (1989) J Microcol Sep 1:294
12. Wieboldt RC, Smith JA (1988) ACS Symposium Series 366:229
13. Shah S, Taylor LT (1989) J High Resolut Chromatogr 12:599
14. Ashraf-Khorassani M, Taylor LT (1989) Anal Chem 61:145
15. Hedrick JL, Calvey EM, Taylor LT (1988) Pittsburgh Conference, New Orleans, LA; February 22–26, paper no 1068
16. Jordan JW, Taylor LT (1986) J Chromatogr Sci 24:82
17. Morin P, Caude M, Rosset R (1987) J Chromatogr 407:87
18. Shafer KH, Pentoney LL Jr, Griffiths PR (1986) Anal Chem 58:58
19. Griffiths PR. In: Ferraro JR, Basile LJ (eds) Fourier Transform Infrared Spectroscopy. Academic Press, New York, Vol 1, 143
20. Johnson CC, Taylor LT (1984) Anal Chem 56:2642
21. Pentoney SL Jr, Shafer KH, Griffiths PR (1986) J Chromatogr Sci 24:230
22. Raymer JH, Moseley MA, Pellizzare ED, Velez GR (1988) J High Resolut Chromatogr Chromatogr Comm 11:209
23. Fuoco R, Pentoney SL Jr, Griffiths PR (1989) Anal Chem 61:2212
24. Fujimoto C, Hirata Y, Jinno K (1985) J Chromatogr 332:47
25. Shafer K, Griffiths PR (1983) Anal Chem 55:1939
26. Raynor MW, Bartle KD, Davis IL, Williams A, Cliffort AA, Chalmers JM, Cook BW (1980) Anal Chem 60:427
27. Raynor MW, Davies IL, Bartle KD, Williams A, Chalmers JM, Cook BW (1987) Eur Chromatogr News 1(4):19
28. Griffiths PR, Pentoney SL, Pariente GL, Norton KL (1988) Mikrochimica Acta Wein, III, 47

11 Supersonic Jet Spectroscopy with Supercritical Fluids

Chung Hang Sin, Steven R. Goates, Milton L. Lee, and David M. Lubman

11.1 Introduction

Selectivity is one of the most important aspects in analytical chemistry. In spectroscopy, selectivity results from high spectral resolution. A molecular spectrum is usually broad due to inhomogeneous broadening and thermal population of higher vibrational and rotational states. The situation becomes even worse for larger molecules, due to closely spaced rotational states and low energy vibrational modes. These broad spectra are not suitable for compound identification.

There are several 'line narrowing' techniques, which can improve the sharpness of the spectra. The most popular ones are matrix isolation [1], the Shpol'skii effect [2], fluorescence line narrowing [3] and the supersonic expansion [4, 5]. Although they operate under different mechanisms, all these techniques serve to lower the temperature of the molecules under investigation. By so doing, mainly the ground state is populated and the spectra now contains only a few sharp peaks. Except for the supersonic expansion, all the other methods use a coolant such as liquid helium to freeze the analyte molecules inside a host matrix. Therefore, there are spectral shifts due to matrix interactions. In a supersonic expansion, sample molecules are seeded into a jet of carrier gas. After a sufficient expansion distance, the molecules are essentially isolated from one another, and thus, are free from any interferences resulting from intermolecular interaction.

As a gas phase technique, the use of supersonic jet spectroscopy (SJS) has been limited to volatile compounds. Recently, two different approaches have been taken to extend supersonic expansion to the analysis of nonvolatile compounds. They are laser desorption [6] and supercritical fluid injection [7, 8]. The first method uses a laser beam to produce a rapid heating effect, so that neutrals and ions can be evaporated into the gas phase without extensive dissociation. The second method makes use of the high solvating power of a dense supercritical fluid to carry the molecules into the jet.

This chapter will focus only on supersonic jet spectroscopy with supercritical fluids. A brief description on the principles of supersonic expansion will be given. Then, practical aspects of supercritical fluid injection will be discussed. Finally, the coupling of supercritical fluid chromatography (SFC) to supersonic jet spectroscopy will be examined.

11.2 Supersonic Jet Spectroscopy

11.2.1 Principles of Supersonic Expansions

Supersonic expansion has been studied for decades in the fields of gas dynamics. It has been recently used in spectroscopy for high resolution spectral analysis. There are quite a number of excellent reviews on the technique itself, and its analytical applications [4, 5, 9–11]. Some of the basic principles of supersonic expansions will be given in this section.

Figure 1 shows a schematic diagram of a supersonic expansion. A gas from a reservoir of pressure, P_0 and temperature, T_0 expands through an orifice of diameter, D, into a region of lower pressure, P_1 and temperature, T_1. The condition of supersonic expansion is met when D is greater than the mean free path of the gas molecules on the high pressure side. Through numerous two-body collisions at the orifice, the random translational motions of the molecules in the reservoir are converted into a directed forward motion. Consequently, a narrowed translational energy distribution is reached, representing a low Maxwell-Boltzmann temperature. The molecules are now supersonically 'cooled'. For a polyatomic molecule, the rotational and vibrational energies are also relaxed into the translationally cold bath. In general, the rotational temperature is cooled to about the same level as the translational temperature. Vibrational temperature is usually higher due to less efficient energy relaxation.

The degree of cooling can be represented by the Mach number, M, which is the ratio of the beam velocity to the local speed of sound in the jet. In general, the cooling process is most efficient when monatomic carrier gases are used, because these gases do not have internal energy. The most commonly used carrier gases are Ar and He. At the orifice, M equals one. As the gas expands further into the vacuum, M increases according to the following equation

$$M = A(X/D)^{\gamma-1} \qquad (1)$$

where A is a constant which depends on γ and is 3.26 for a monatomic gas. At a certain distance from the orifice, the jet density becomes so low that

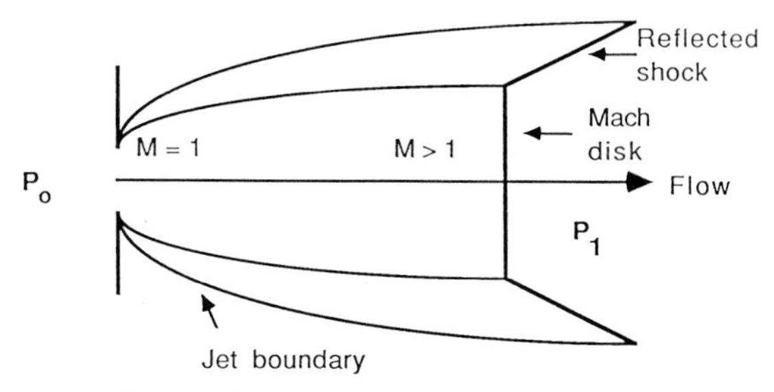

Fig. 1. Schematic diagram of a supersonic expansion

the molecules are essentially isolated from one another. At this point, the cooling process stops, because there are no more collisions. The molecules now enter a free flow region and the Mach number reaches a maximum value, known as the terminal Mach number, M_T. For a monatomic gas such as Ar,

$$M_T = 133\,(P_0 D)^{0.4} \quad . \tag{2}$$

For helium, because of quantum effects, which cause the collisional cross section to increase with decreasing relative energy of the colliding atoms, the terminal Mach number is much larger than the value predicted by this equation. Upon substituting the calculated M_T into Eq. (1), one would be able to determine where the free flow region starts.

The Mach disk is the boundary at the end of the jet, where the expanding gas molecules merge with the background molecules. The distance of the Mach disk from the orifice, X_m, can be calculated by the following equation.

$$X_m = 0.667 \times (P_0/P_1)^{1/2} \quad . \tag{3}$$

The Mach disk should be located far beyond where the free flow region begins, so that maximum cooling can be assured. If P_1 is low enough, X_m may actually be longer than the dimension of the vacuum chamber. This simply means that the jet is well preserved throughout the chamber. Thus, Eq. (3) is useful for determining the necessary pumping requirements.

11.2.2 Spectroscopic Methods for Detection

The two most popular methods used for supersonic jet spectroscopy are laser induced fluorescence (LIF) and resonant two-photon ionization (R2PI) [9]. In LIF, an excitation spectrum is obtained when the excitation wavelength is scanned while the total fluorescence signal is collected. A dispersed fluorescence spectrum is obtained when the excitation wavelength is fixed and the emitted light is dispersed and collected as a function of wavelength. Both methods provide high resolution spectra with a supersonic jet. There are two major disadvantages of fluorescence. Firstly, the method is limited to molecules of high fluorescence quantum efficiency. Secondly, the method's detection limit is affected by the presence of scattered light.

These two disadvantages of LIF do not exist in R2PI. Under an intense laser beam, a molecule can absorb two photons 'simultaneously'. With the right wavelength, the first photon excites the molecule to a real intermediate state, and the second photon brings the molecule above its ionization continuum and produces a molecular ion. The detection of ions is not subjected to interference from scattered light. In addition, ions produced by R2PI can be separated in a mass spectrometer for one more dimension of information.

Other detection methods include nonlinear Raman [12], absorption [13], bolometric detection [14], phosphorescence [15] and infrared fluorescence [16].

In the following sections, the discussion will be focussed on the use of supercritical fluids and subcritical fluids for introducing nonvolatile compounds into a supersonic jet. Then, the interfacing of supercritical fluid chromatography to supersonic jet spectroscopy will be presented.

11.3 High Pressure Fluid Injection for Nonvolatile Samples

As a gas phase technique, supersonic jet spectroscopy is limited to compounds, which have sufficient vapor pressure. Although, heating might be used to increase the vapor pressure, it is not suitable for thermally labile compounds. The use of supercritical fluids provides an improved mean for volatilizing nonvolatile compounds into the jet. Many nonvolatile compounds have been found to be soluble in supercritical CO_2 and subsequently expanded into the mass spectrometer [7, 17]. Because a supercritical fluid has a high density, it has enhanced solvating power. In the case of polar molecules, more polar fluids, such as supercritical NH_3, and high pressure methanol and water were used [23].

11.3.1 Experimental

Figure 2 shows the experimental setup for high-pressure fluid injection with R2PI detection. The chamber was pumped differentially using a large skimmer (2.4 cm diameter) on the partition plate. A pulsed solenoid valve synchronized with the laser introduced the high pressure fluid into the vacuum. After passing through the skimmer, the sample molecules were intercepted by a laser beam perpendicular to the molecular beam axis. There were four liquid nitrogen traps extending from the top of the chamber into the vacuum. By condensing the fluids expanded from the valve, the pressure of the first and the second compartments remained below 1×10^{-4} Torr and 2×10^{-5} Torr respectively.

The most important feature in this experiment was the high pressure pulsed valve [18]. This valve could withstand a pressure up to 400 atm. The valve seals via a metal-to-metal contact. The first version of this valve was designed with the solenoid enclosed inside the valve and was not suitable for corrosive liquids such as ammonia. In the version currently in use, the solenoid was wound outside the valve itself. This arrangement made it possible to use ammonia as the carrier fluid [9]. The gas pulse was about

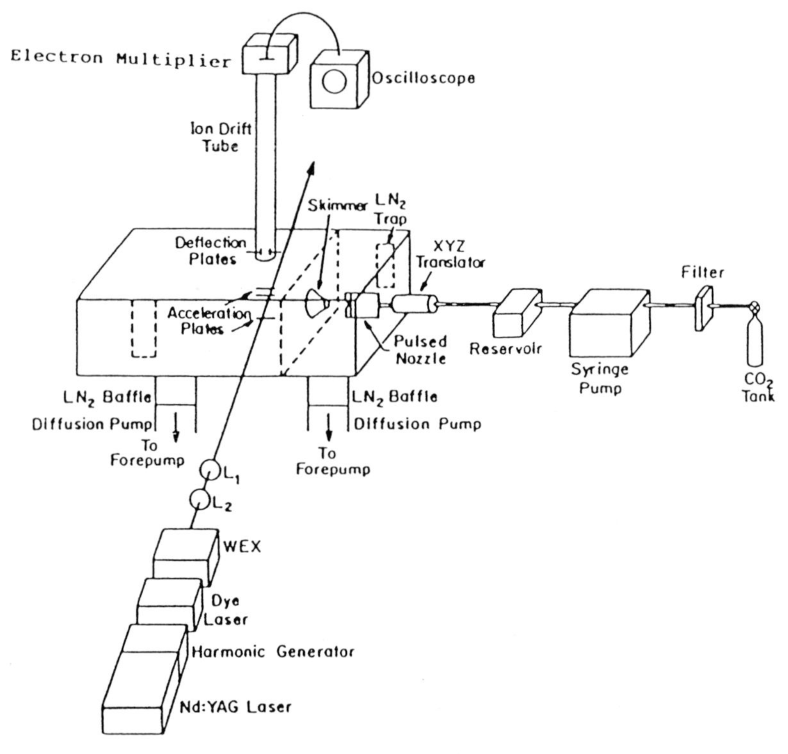

Fig. 2. Experimental setup for supercritical fluid injection into supersonic jet. Reprinted with permission from Fig. 1 in Ref. [7]. Copyright American Chemical Society

200 μs FWHM. The use of this pulsed nozzle allowed us to use a large orifice (200 μm) as compared to that possible in a continuous flow.

R2PI was used to produce ions from the supersonic molecular beam. The ionization source was a Spectra Physics frequency-doubled Nd:YAG pumped dye laser. In the case of single wavelength ionization, the fourth harmonic (266 nm) was used. The laser power was about 2×10^6 W/cm^2. The mass spectra were obtained by averaging the signal over 100 laser shots with a LeCroy 9400 transient digitizer. In order to produce a wavelength ionization spectrum, a boxcar integrator was used to monitor the peak intensity of the molecular ion, while the wavelength was scanned. A Varian 8500 syringe pump was used to provide the required high pressure. The nozzle was usually kept below 100 °C except for water and ammonia which required slightly higher temperature to prevent clustering.

11.3.2 Supercritical Fluid Injection

Supercritical CO_2 and N_2O were used to carry nonpolar compounds into the jet. Figure 3 shows a wavelength ionization spectrum of carbazole using

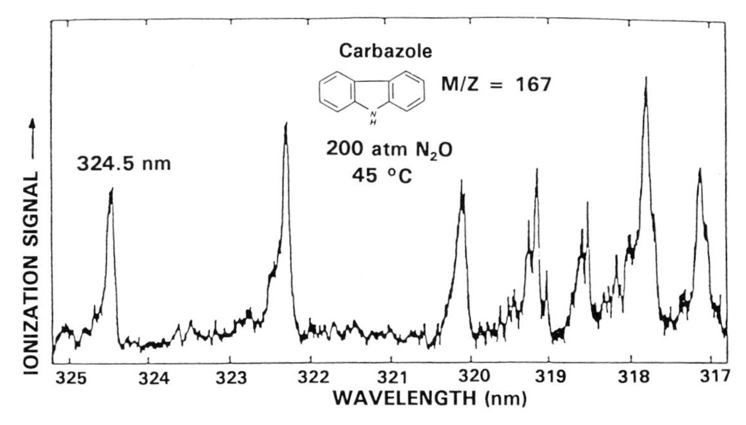

Fig. 3. R2PI wavelength spectrum of carbazole expanded in a supersonic jet from supercritical N_2O at 200 atm and 45 °C

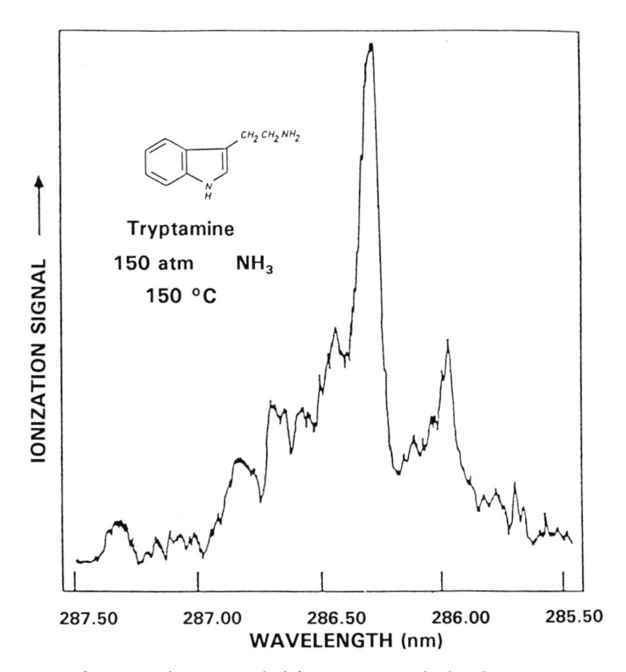

Fig. 4. R2PI wavelength spectrum of tryptamine expanded in a supersonic jet from supercritical NH_3 at 150 atm and 150 °C. Reprinted with permission from Fig. 2 in Ref. [21]. Copyright American Chemical Society

200 atm N_2O injection at 45 °C. If 1 atm of gas is used, the temperature would need to be at least 200 °C before sufficient signal could be detected. This clearly demonstrates the usefulness of supercritical fluids for carrying nonvolatile samples into the supersonic jet at mild temperatures. CO_2 and

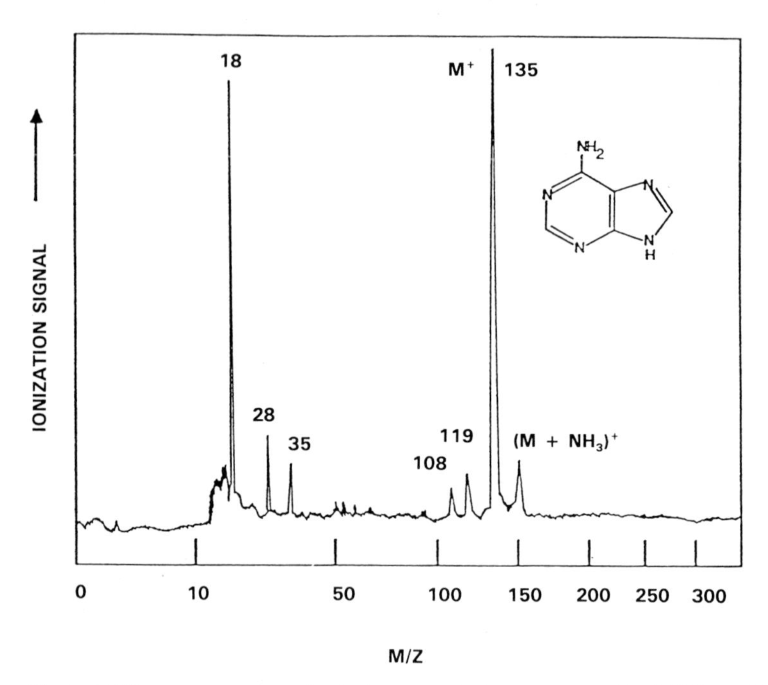

Fig. 5. R2PI mass spectrum of adenine expanded from supercritical NH_3 at 120 atm and 150 °C. Reprinted with permission from Fig. 3 in Ref. [19]. Copyright American Institute of Physics

N_2O are generally useful for nonpolar compounds, such as PAH's. For polar molecules, derivatization has been used to increase solubility in supercritical CO_2 and N_2O [20].

Another means of increasing solubility is to use supercritical ammonia as the carrier [21]. Figure 4 shows the wavelength spectrum of tryptamine injected by supercritical NH_3 at 150 atm and 150 °C. With the increase in polarity, the stronger solute-solvent interaction not only aids the solvation of the polar molecules, but also increases cluster formation as seen in Fig. 5. However, the degree of clustering is still low. It can further be minimized by adjusting the delay of the laser so that it only probes the leading edge of the gas pulse, where the beam density is low enough to prevent extensive clustering [22]. However, because of its toxicity, ammonia may not be considered an ideal carrier gas for supersonic jet injection.

11.3.3 High Pressure Liquid Injection

Methanol and water are very good solents for polar compounds. Because the critical temperatures for methanol and water are very high, they are used under their subcritical conditions (i.e. at a lower temperature) in these ex-

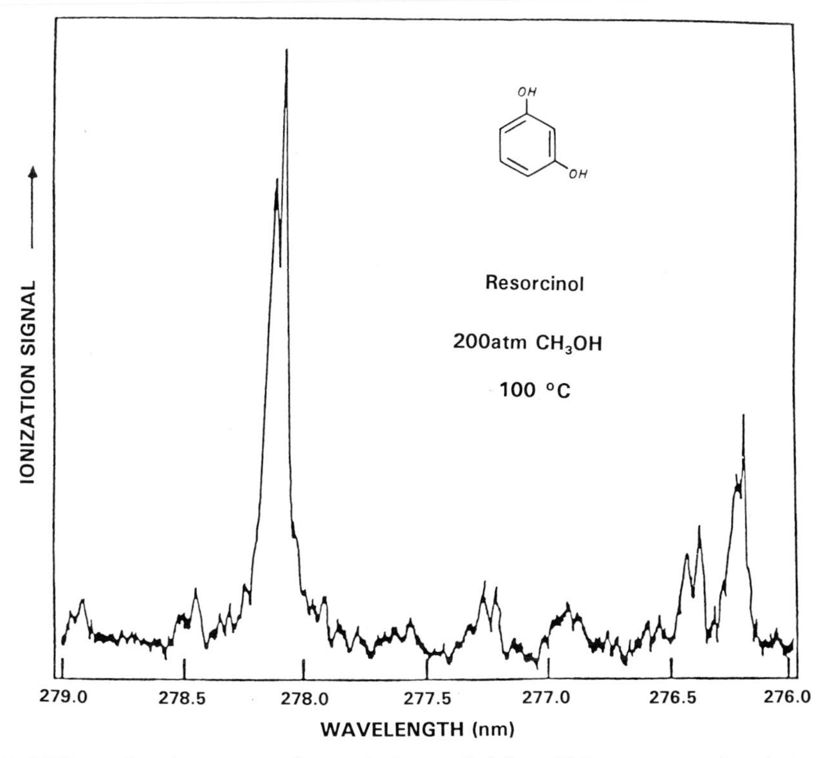

Fig. 6. R2PI wavelength spectrum of resorcinol expanded from high pressure methanol at 200 atm and 100 °C. Reprinted with permission from Fig. 2(a) in Ref. [23]. Copyright Society for Applied Spectroscopy

periments. Figure 6 shows the wavelength spectrum of resorcinol using methanol at 200 atm and 100 °C. Because methanol has a much higher internal heat capacity than CO_2 and Ar, the spectral peak is expected to be broader. The spectral resolution is, however, still relatively high. Numerous catecholamine metabolites and neuroleptic drugs have been introduced into the jet using high pressure methanol and water [23].

11.4 Supercritical Fluid Chromatography and Supersonic Jet Spectroscopy

In addition to solvating power, supercritical fluids have good diffusivities. These two characteristics make supercritical fluids very good candidates for the mobile phase in chromatography. As mentioned in a previous section, nonpolar compounds can be introduced into a supersonic jet by using a supercritical fluid as a carrier. Thus, it is possible to use supersonic jet spectroscopy as a selective detection method for supercritical fluid chromatography [24, 25]. This is especially useful for the analysis of complex mixtures, which often require more than one dimension of selectivity.

11.4.1 Experimental

Figure 7 shows the experimental setup for SFC-SJS. Pressure was provided by a modified ISCO 314 syringe pump. Pressure programming was controlled by a home-built interface. A Valco injection valve with an internal volume of 0.1 or 0.2 μl was used. In capillary SFC, a split line was used to limit the load on the column. The column was housed inside a Hewlett-Packard 5710A GC oven. The interfacing assembly was held on a X-Y-Z positioning stage.

In this experiment, laser-induced fluorescence was used to detect the eluents in the jet. The excitation source was a Lambda Physik excimer pumped dye laser. The laser beam was perpendicular to the molecular beam and the total fluorescence was collected at right angles to the laser beam and the molecular beam, and focussed into an monochromator or directly

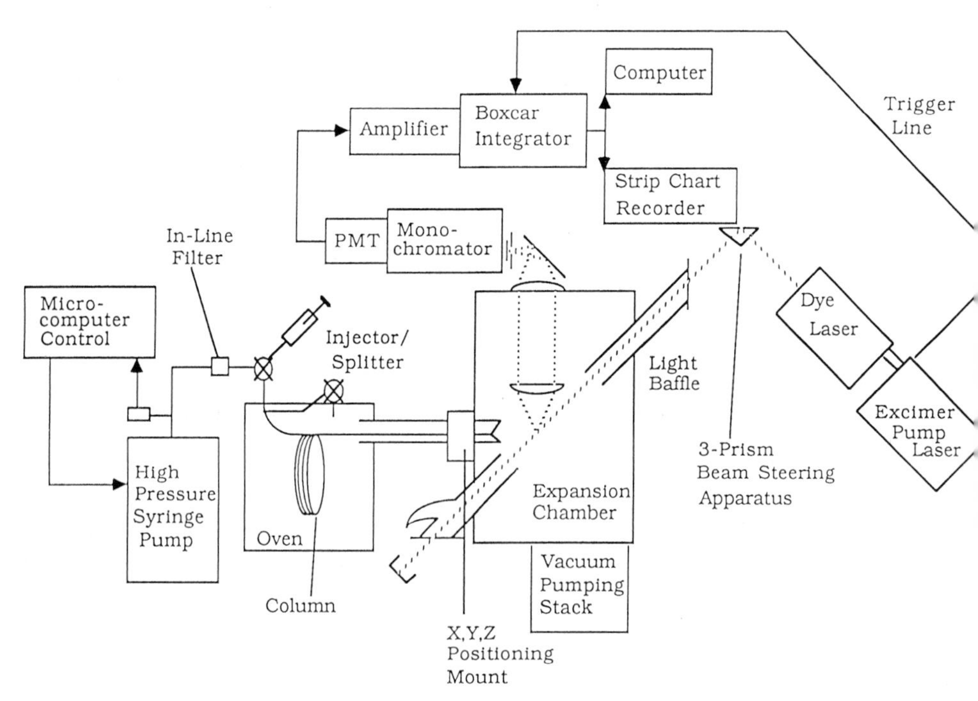

Fig. 7. Experimental setup for supercritical fluid chromatography/supersonic jet spectroscopy. Reprinted with permission from Fig. 1 in Ref. [24]. Copyright Aster Publishing Corporation

into a photomultiplier tube. Selective detection of perylene in coal tar was performed by setting the excitation wavelength at 415.5 nm, which is the 0-0 transition of perylene.

11.4.2 The Interfaces

Two kinds of interfaces were used here. The direct expansion nozzle as shown in Fig. 8 is the most straightforward approach. The 50 μm capillary transfer line was restricted by direct coupling to a pinhole orifice. Since light scattering would be a problem for LIF, the nozzle was heated up to 200 °C in order to avoid aerosol and cluster formation. In this continuous flow situation, a small orifice (5 μm) was used to maintain the column pressure. Nevertheless, the flow rate would still be too high for a capillary separation. Figure 9 shows a selective detection of perylene in a coal tar sample separated in a 25 cm×1 mm i.d., C_8 packed column [24]. As indicated in the figure, pressure programming was used.

For capillary SFC, because of very low flow rate, a supersonic jet cannot be produced directly from the column effluent. A sheath-flow nozzle (Fig. 10) was used instead. In this nozzle, the capillary transfer line was prerestricted by a porous frit or an integral restrictor at the end. Ar at 6–8 atm was used as a make-up gas. As a result, a larger orifice (25 μm) could be used. Since the eluents from the column were swept out of the nozzle immediately by the make-up gas, the effective dead volume was negligible. As expected, the Ar make-up gas provided a better cooling efficiency for the seeded molecules coming out from the capillary. The linewidth was about 0.04 nm, which is half of the value obtained with the direct expansion nozzle. The use of a make-up gas also reduced cluster formation, and made

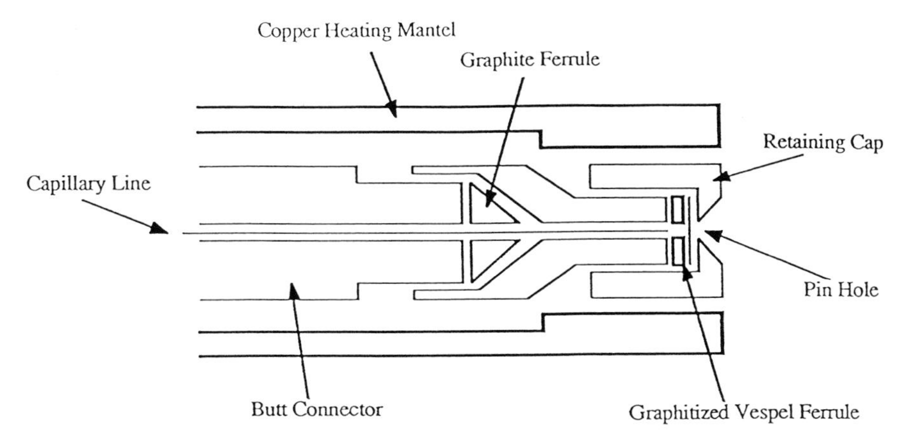

Fig. 9. Selective supersonic jet detection of perylene in a coal tar with micro-bore, packed SFC. Excitation wavelength: 415.5 nm; total fluorescence

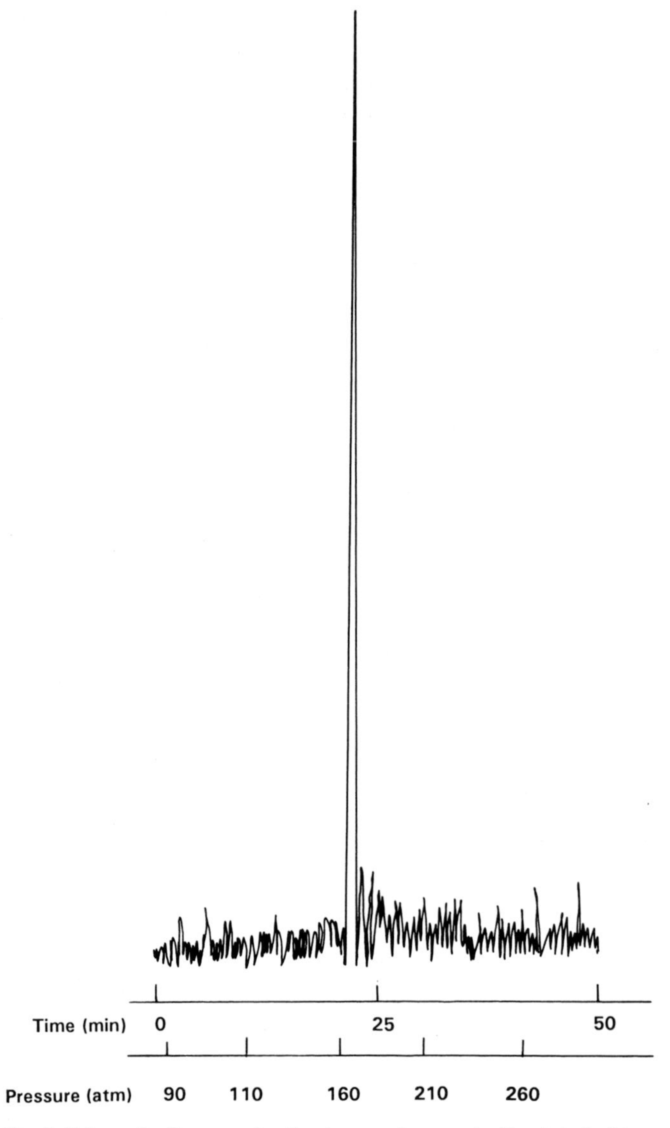

Fig. 8. Schematic diagram of a direct expansion nozzle. Reprinted with permission from Fig. 2 in Ref. [24]. Copyright Aster Publishing Corporation

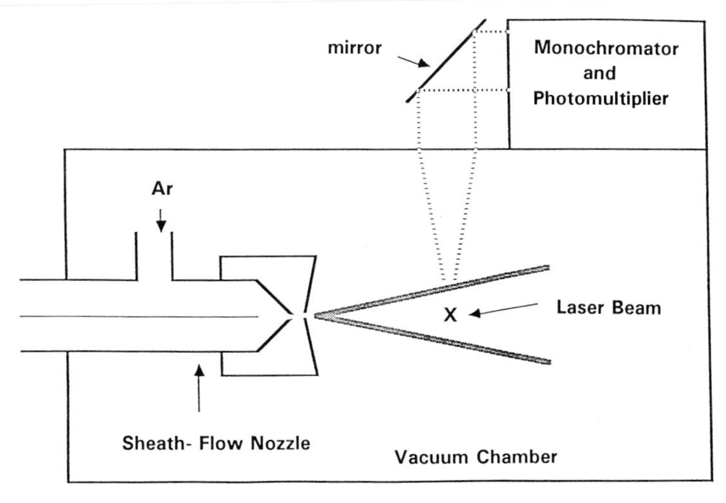

Fig. 10. Schematic diagram of a sheath-flow nozzle used in supercritical fluid chromatography/supersonic jet spectroscopy

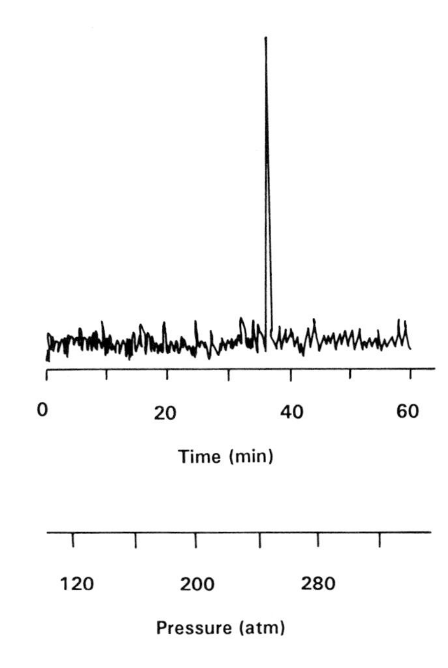

Fig. 11. Selective supersonic jet detection of perylene in a coal tar with capillary SFC. Excitation wavelength: 415.5 nm; total fluorescence. Reprinted with permission from Fig. 4(b) in Ref. [25]. Copyright Aster Publishing Corporation

it possible to run the nozzle at a lower temperature (e.g. 160 °C). Figure 11 shows a selective detection of perylene in a coal tar sample separated in a 5 m × 50 μm i.d. fused-silica capillary, coated with a 0.25 μm film thickness of an *n*-octylpolysiloxane stationary phase [25].

11.5 Conclusions

It has been shown that supercritical and subcritical high pressure fluids can be used to inject nonvolatile compounds into a supersonic jet. For nonpolar compounds, supercritical CO_2 and N_2O can be used as carrier fluids. For polar compounds, supercritical ammonia, subcritical methanol and water are useful. It has also been demonstrated that supercritical fluid chromatography can be coupled to supersonic jet spectroscopy by two kinds of interfaces. A direct expansion nozzle is suitable for packed SFC, which has a higher flow rate. A sheath-flow nozzle is needed for capillary SFC to avoid flow rate mismatch between the two techniques.

References

1. Wehry EL, Mamantov G (1979) Anal Chem 51:643A
2. D'Silva AP, Fassel VA (1984) Anal Chem 56:985A
3. Heisig V, Jeffrey AM, McGlade MJ, Small GJ (1984) Science 223:289
4. Levy DH (1980) Annu Ref Phys Chem 31:197
5. Smalley RE, Wharton L, Levy DH (1977) Acc Chem Res 10:139
6. Li L, Lubman DM (1988) Anal Chem 60:2591
7. Sin CH, Pang HM, Lubman DM, Zorn J (1986) Anal Chem 58:487
8. Goates SR, Barker AJ, Zakharia HS, Khoobehi B, Sheen CW (1987) Appl Spectrosc 41:1392
9. Lubman DM (1987) Anal Chem 59:31A
10. Hayes JM (1987) Chem Rev 87:745
11. Goates SR, Sin CH (1989) Appl Spectrosc Rev 25(2):81
12. Huber-Walchli P, Nibler JW (1982) J Chem Phys 76:273
13. Amirav A, Even U, Jortner J (1982) Anal Chem 54:1666
14. Miller RE (1982) Rev Sci Instrum 53:1719
15. Abe H, Kamei S, Mikami N, Ito M (1984) Chem Phys Lett 109:217
16. Stewart GM, Ensminger MD, Kulp TJ, Ruoff RS, McDonald JD (1983) J Chem Phys 79:3190
17. Smith RD, Udseth HR (1984) Anal Chem 55:2266
18. Pang HM, Sin CH, Lubman DM, Zorn J (1986) Anal Chem 58:1581
19. Pang HM, Lubman DM (1988) Rev Sci Instrum 59:2460
20. Pang HM, Sin CH, Lubman DM (1988) Spectrochim Acta Part B 43:671
21. Pang HM, Lubman DM (1989) Anal Chem 61:777
22. Tembreull R, Sin CH, Pang HM, Lubman DM (1985) Anal Chem 57:2911
23. Pang HM, Sin CH, Lubman DM (1988) Appl Spectrosc 42:1200
24. Simon JK, Sin CH, Zabriskie NA, Lee ML, Goates SR, Fields SM (1989) J MicroCol Sep 1:200
25. Goates SR, Sin CH, Simons JK, Markides KE, Lee ML (1989) J MicroCol Sep 1:207

Basic References Analytical Scale SFC/SFE

Books and Journals

Lee ML, Markides KE (1990) Analytical Supercritical Fluid Chromatography and Extraction. Chromatography Conferences Inc Provo, Utah

Yoshioka M, Parvez S, Miazaki T, Parvez H (1989) Supercritical Fluid Chromatography and Micro-HPLC Progress in HPLC. VSP, Utrecht

Smith RM (1988) Supercritical Fluid Chromatography. RSC Chromatography Monographs

White CM (ed) (1988) Modern Supercritical Fluid Chromatography Chromatographic Methods. Dr. Alfred Hüthig Verlag, Heidelberg

Charpentier BA, Sevenants MR (1988) Supercritical Fluid Extraction and Chromatography, Techniques and Applications. ACS Symposium Series 406

Stahl E, Quirin KW, Gerard D (1987) Verdichtete Gase zur Extraktion und Raffination. Springer-Verlag, Berlin Heidelberg New York

Introductory Journals

Angewandte Chemie intern eng edit 1978, 10, whole issue

General Journals

Analytical Chemistry
Fresenius J Anal Chem

Chromatography and Extraction Journals

J Chromatogr
J High Res Chrom
Chromatographia
J Microcolumn Sep
J Supercritical Fluids

Theoretical Journals

Ber Bunsenges Phys Chem
Fluid Phase Equilibria
J Chem Eng Data
J Chem Thermodynamics
J Solution Chemistry

Reviews

SFE

Hawthorne SB (1990) Analytical-scale supercritical fluid extraction. Anal Chem 62:633A
Vannoort RW, Chervet JP, Lingeman H, Dejong GJ, Brinkman UATh (1990) Coupling of supercritical fluid extraction with chromatographic techniques. J Chromatography 505:45

SFC

Janssen HG, Cramers CA (1990) Some aspects of capillary supercritical fluid chromatography. J Chromatography 505:19
Klesper E, Schmitz FP (1988) Gradient Methods in Supercritical Fluid Chromatography. J Supercritical Fluids 1:45

Applications and Proceedings

McHugh M (ed) (1991) 2nd International Symposium on Supercritical Fluids
McHugh M (1991) Dept of Chem Eng, John Hopkins University, Baltimore, MD 21218. International Symposium on Supercritical Fluid Chromatography and Extractions, Abstracts. Park City, Utah
Markides KE (ed) (1991) European Symposium on Analytical SFC and SFE. Wiesbaden, December 4–5
Perrut M (1988) International Symposium on Supercritical Fluids, Tome 1+2. Société française de chimie, october 17–19, Nice
SFC Applications (1988) Workshop, Park City, Utah
SFC Applications (1989) Workshop, Snow Bird, Utah

Instrumentation

Köhler J, Rose A, Schomburg G (1988) Instrumentation for SFC Systems: Different Sampling and Restriction Designs. Journal of High Resolution Chromatography & Chromatography Communications 11:191

Injection

Tuominen JP, Markides KE, Lee ML (1991) Optimization of Internal Valve Injection in Open Tubular Column Supercritical Fluid Chromatography. J Microcol 3:229
Lee ML, Xu B, Huang EC, Djordjevic NM, Chang HCK, Markides KE (1989) Liquid Sample Introduction Methods in Capillary Column Supercritical Fluid Chromatography. J Microcolumn 1:7

Detection

Webster GK, Carnahan JW (1992) Atomic Emission Detection for Supercritical Fluid Chromatography Using a Moderate-Power Helium Microwave-Induced Plasma. Anal Chem 64:50

Luffer DR, Novotny MV (1991) Capillary Supercritical Fluid Chromatography and Microwave-Induced Plasma Detection of Cyclic Boronate Esters of Hydroxy Compounds. J Microcol 3:39

Munder A, Christensen RG, Wise StA (1991) Microanalysis of Explosives and Propellants by On-Line Supercritical Fluid Extraction/Chromatography with Triple Detection. J Microcol 3:127

Howard AL, Taylor LT (1991) Ozone-Based Sulfur Chemiluminescence Detection: Its Applicability to Gas. Supercritical Fluid, and High Performance Liquid Chromatography. Journal of High Resolution Chromatography 14:785

Huang MX, Markides KE, Lee ML (1991) Evaluation of an Ion Mobility Detector for Supercritical Fluid Chromatography with Solvent-Modified Carbon Dioxide Mobile Phases. Chromatographia 31:163

Fields StM, Grolimund K (1989) Evaluation of supercritical Hexafluoride as a mobile phase for polar and non-polar compounds. J Chromatography 472:197

Bornhop DJ, Schmidt St, Porter NL (1988) Use of simultaneous detectors in capillary supercritical fluid chromatography. J Chromatography 459:193

Later DW, Bornhop DJ, Lee ED, Henion JD, Wieboldt RC (1987) Detection techniques for capillary supercritical Fluid chromatography. LC GC 9:804

Eatherton RL, Morrissey MA, Siems WF, Hill HH (1986) Ion Mobility Detection after Supercritical Fluid Chromatography. J High Res Chromatogr & Chromatogr Comm 9:154

Novotny M (1986) New Detection Strategies Through Supercritical Fluid Chromatography. J High Res Chromatogr & Chomatogr Comm 9:137

Johnson CC, Jordan JW, Taylor LT (1985) On-Line Supercritical Fluid Chromatography with Fourier Transform Infrared Spectrometric Detection Employing Packed Columns and a High Pressure Lightpipe Flow Cell. Chromatographia 20:717

Fjeldsted JC, Richter BE, Jackson WP, Lee ML (1983) Scanning Fluorescence Detection in Capillary Supercritical Fluid Chromatography. J Chromatography 279:423

Mobile/Stationary Phases

Malik A, Jinno K (1991) Retention Behaviour of Aromatic Compounds in Liquid Chromatography and Supercritical Fluid Chromatography with Coarse-Particles of Bonded β-Cyclodextrin Stationary Phase. Chromatographia 31:561

Röder W, Ruffing F-J, Schomburg G, Pirkle WH (1987) Chiral SFC-Separations Using Polymer-Coated Open Tubular Fused Silica Columns Comparison of Enantiomeric Selectivity in SFC and LC Using the Same Stationary Phase of the Pirkle Type. J HRCC 10:665

Mourier P, Sassiat M, Caude M, Rosset R (1986) Retention and Selectivity in Carbon Dioxide Supercritical Fluid Chromatography with Various Stationary Phases. J Chromatography 353:61

Leren E, Landmark KE, Greibrokk T (1991) Sulphur Dioxide as a Mobile Phase in Supercritical Fluid Chromatography. Chromatographia 31:535

Berger TA, Deye JF (1991) Efficiency in Packed Column Supercritical Fluid Chromatography Using a Modified Mobile Phase. Chromatographia 31:519

Crow JA, Foley JP (1991) Formic Acid Modified Carbon Dioxide as a Mobile Phase in Capillary Supercritical Fluid Chromatography. J Microcol Sep 3:47–57

Lochmüller CH, Mink LP (1990) Comparison of supercritical carbon dioxide and supercritical propane as mobile phases in supercritical fluid chromatography. J Chromatography 505:119

Raynor MW, Shilstone GF, Clifford AA, Bartle KD, Cleary M, Cook BW (1991) Xenon as a Mobile Phase in Supercritical Fluid Chromatography. J Microcol Sep 3:337
Ong ChP, Lee HE, Yau Li SF (1990) Chlorodifluoromethane as the Mobile Phase in Supercritical Fluid Chromatography of Selected Phenols. Anal Chem 62:1389
Geiser FO, Yocklovich SG, Lurcott SM, Guthrie JW, Levy EJ (1988) Water as a stationary phase modifier in packed column supercritical fluid chromatography. J Chromatography 459:173
Kuel JC, Markides KE, Lee ML (1987) Supercritical Ammonia as Mobile Phase in Capillary Chromatography. JHRCC 10:257
Yonker CR, Smith RD (1986) Studies of Retention Processes in Capillary Supercritical Fluid Chromatography with Binary Fluid Mobile Phases. J Chromatography 361:25

Recommended Literature

SFC 1992

Vérillon F, Heems D, Pichon B, Coleman K, Robert JC (1992) Supercritical fluid chromatography – With independent programming of mobile- phase pressure, composition and flow rate. International Laboratory, July/Aug., 29

Jagota NK, Stewart JT (1992) Separation of non-steroidal anti-inflammatory agents using supercritical fluid chromatography. J Chromatogr 604:255

Zhang X, Martire DE (1992) Characterization of a supercritical fluid chromatographic retention process with a large pressure drop by the temporal average density. J Chromatogr 603:193

Giddings LD, Okesik SV, Pekay LA (1992) Retention characteristics of high-molecular-weight compounds in capillary supercritical fluid chromatography. J Chromatogr 603:205

Pasch H, Krüger H, Much H (1992) Analysis of benzyloxy-terminated poly(1,3,6-trioxocane)s by supercritical fluid chromatography. J Chromatogr 589:295

Laintz KE, Shieh GM, Wai CM (1992) Simultaneous Determination of Arsenic and Antimony Species in Environmental Samples using Bis(trifluoroethyl)dithiocarbamate Chelation and Supercritical Fluid Chromatography. J Chromatographic Science 30:120

Janssen HG, Snijders H, Cramers C, Schoenmakers P (1992) Compressibility Effects in Packed and Open Tubular Gas and Supercritical Fluid Chromatography. JHRC 15:458

Thomson JS, Rynaski AF (1992) Simulated Distillation of Wax Samples Using Supercritical Fluid and High Temperature Gas Chromatography. JHRC 15:227

Jones BA, Shaw TJ, Clark J (1992) The Effect of Temperature on Selectivity in Capillary SFC. J Microcol Sep 4:215

Injection

Koski IJ, Markides KE, Richter BE, Lee ML (1992) Microliter Sample Introduction for Open Tubular Column Supercritical Fluid Chromatography Using a Packed Capillary for Solute Focusing. Anal Chem 64:1669

Oudsema JW, Poole CF (1992) Some Practical Experiences in the Use of a Solventless Injection System for Packed Column Supercritical Fluid Chromatography. JHRC 15:65

Cortes HJ, Campbell RM, Himes RP, Pfeiffer CD (1992) On-Line Coupled Liquid Chromatography and Capillary Supercritical Fluid Chromatography: Large-Volume Injection System for Capillary SFC. J Microcol Sep 4:239

Berg BE, Flaaten AM, Paus J, Greibrokk T (1992) Extended Use of Solvent Venting Injection Techniques for Large Sample Volumes and Coupled Capillary Columns in SFC. J Microcol Sep 4:227

Stationary/Mobile Phases

Berger TA, Deye JF (1992) Correlation between column surface area and retention of polar solutes in packed- column supercritical fluid chromatography. J Chromatogr 594:291

Johnson DF, Bradshaw JS, Eguchi M, Rossiter BE, Lee ML (1992) Synthesis of (1R-trans)-N,N′-1,2-cyclohexylenebisbenzamideoligodimethylsiloxane copolymers for use as chiral stationary phases for capillary supercritical fluid chromatography. J Chromatogr 594:283

Petersson P, Markides KE, Johnson DJ, Rossiter BE, Bradshaw JS, Lee ML (1992) Chromatographic Evaluation of Chiral (1R-trans)-N,N′-1,2- Cyclohexylenebisbenzamide-oligodimethylsiloxane Copolymeric Stationary Phases for Capillary Supercritical Fluid Chromatography. J Microcol Sep 4:155

Page StH, Sumpter SR, Lee ML (1992) Fluid Phase Equilibria in Supercritical Fluid Chromatography with CO2-Based Mixed Mobile Phases: A Review. J Microcol Sep 4:91

Detection

Pinkston JD, Delaney TE, Morand KL, Cooks RG (1992) Supercritical Fluid Chromatography/Mass Spectrometry Using a Quadrupole Mass Filter/Quadrupole Ion Trap Hybrid Mass Spectrometer with External Ion Source. Anal Chem 64:1571

Webster GK, Carnahan JW (1992) Atomic Emission Detection for Supercritical Fluid Chromatography Using a Moderate-Power Helium Microwave-Induced Plasma. Anal Chem 64:50

Sye WF, Zhao ZX, Lee ML (1992) Comparative Application of Sulfur Chemiluminescence Detection in Gas and Supercritical Fluid Chromatography. Chromatographia 33:507

Upnmoor D, Brunner G (1992) Packed Column Supercritical Fluid Chromatography with Light-Scattering Detection. II. Retention Behavior of Squalane and Glucose with Mixed Mobile Phases. Chromatographia 33:261

Upnmoor D, Brunner G (1992) Packed Column Supercritical Fluid Chromatography with Light-Scattering Detection. I. Optimization of Parameters with a Carbon Dioxide / Methanol Mobile Phase. Chromatographia 33:255

Ashraf-Khorassani M, Levy JM (1992) Evaluation of thermionic detector for packed capillary supercritical fluid chromatography of nitrogen-containing compounds. Fresenius J Anal Chem 342:688

Yang Y, Baumann W (1992) Chromatography with dense gases – I. On the sensitivity of a flame ionisation detector in a supercritical fluid chromatograph. Fresenius J Anal Chem 342:684

Morin-Allory L, Berbreteau B (1992) High-performance liquid chromatography and supercritical fluid chromatography of monosaccharides and polyols using light-scattering detection. J Chromatogr 590:203

Hill HH, St Louis RH, Morrissey MA, Shumate CB, Siems WF, McMinn DG (1992) A Detection Method for Unified Chromatography: Ion Mobility Monitoring. JHRC 15:417

Lafosse M, Elfakir C, Morin-Allory L, Dreux M (1992) The Advantages of Evaporative Light Scattering Detection in Pharmaceutical Analysis by High Performance Liquid Chromatography and Supercritical Fluid Chromatography. JHRC 15:312

SFE 1992

Hawthorne StB, Langenfeld JJ, Miller DJ, Burford MD (1992) Comparison of Supercritical CHClF2, N2O, and CO2 for the Extraction of Polychlorinated Biphenyls and Polycyclic Aromatic Hydrocarbons. Anal Chem 64:1614

Thomson CA, Chesney DJ (1992) Supercritical Carbon Dioxide Extraction of 2,4-Dichlorophenol from Food Crop Tissues. Anal Chem 64:846

Hawthorne StB, Miller DJ, Nivens DE, White DC (1992) Supercritical Fluid Extraction of Polar Analytes Using in Situ Chemical Derivatization. Anal Chem 64:405

Alexanddrou N, Lawrence MJ, Pawliszyn J (1992) Cleanup of Complex Organic Mixtures Using Supercritical Fluids and Selective Adsorbents. Anal Chem 64:301

Küppers St (1992) The Use of Temperature Variation in Supercritical Fluid Extraction of Polymers for the Selective Extraction of Low Molecular Weight Components from Poly (Ethylene Terephthalate). Chromatographia 33:434

King JW, Hopper ML (1992) Analytical Supercritical Fluid Extraction: Current Trends and Future Vistas. Journal of AOAC International, 75:375

Smith RM, Burford MD (1992) Optimization of supercritical fluid extraction of volatile constituents from a model plant matrix. J Chromatogr 600:175

Lee H-B, Peart TE (1992) Supercritical carbon dioxide extraction of resin and fatty acids from sediments at pulp mill sites. J Chromatogr 594:309

Langenfeld JJ, Burford MD, Hawthorne StB, Miller DJ (1992) Effects of collection solvent parameters and extraction cell geometry on supercritical fluid extraction efficiencies. J Chromatogr 594:297

Messer DC, Taylor LT (1992) Development of Analytical SFE of a Polar Drug from an Animal Food Matrix. JHRC, 15:238

Slack GC, McNair HM, Wasserzug L (1992) Characterization of Semtex by Supercritical Fluid Extraction and Off-Line GC-ECD and GC-MS. JHRC 15:102

Liu Z, Farnsworth PB, Lee ML (1992) High-Speed, Thermally Modulated SFE/GC for the Analysis of Volatile Organic Compounds in Solid Matrices. J Microcol Sep 4:199

King JW, List GR, Johnson JH (1992) Supercritical Carbon Dioxide Extraction of Spent Bleaching Clays. Journal of Supercritical Fluids 5:38

Maxwell RJ, Hampson JW, Cygnarowicz-Provost (1992) Comparison of the Solubility in Supercritical Fluids of the Polycyclic Ether Antibiotics: Lasalocid, Monensin, Narasin, and Salinomycin. J of Supercritical Fluids 5:31

Cygnarowicz-Provost M, O'Brien DJ, Maxwell RJ, Hampson JW (1992) Supercritical-Fluid Extraction of Fungal Lipids Using Mixed Solvents: Experiment and Modeling. J of Supercritical Fluids 5:24

Raghuram GV, Srinivas P, Sastry SVGK, Mukhopadhyay M (1992) Modeling Solute-Co-solvent Interactions for Supercritical-Fluid Extraction of Fragrances. J of Supercritical Fluids 5:19

Bulley NR, Labay L, Arntfield SD (1992) Extraction/Fractionation of Egg Yolk Using Supercritical CO2 and Alcohol Entrainers. J of Supercritical Fluids 5:13

Li S, Hartland St (1992) Influence of Co-Solvents on Solubility and Selectivity in Extraction of Xanthines and Cocoa Butter from Cocoa Beans with Supercritical CO2. J of Supercritical Fluids 5:7

Jennings DW, Deutsch HM, Zalkow LH, Teja AS (1992) Supercritical Extraction of Taxol from the Bark of Taxus Brevifolia. J of Supercritical Fluids 5:1

Others

Sin CH, Linford MR, Goates StR (1992) Supercritical Fluid/Supersonic Jet Spectroscopy with a Sheath-Flow Nozzle. Anal Chem 64:233

Dahl S, Dunalewicz A, Fredenslund A, Rasmussen P (1992) The MHV2 Model: Prediction of Phase Equilibria at Sub- and Supercritical Conditions. J of Supercritical Fluids 5:42

Betts TA, Zagrobelny JA, Bright FV (1992) Investigation of Solute-Fluid Interactions in Supercritical CF3H: A Multifrequency Phase and Modulation Fluorescence Study. J of Supercritical Fluids 5:48

Hamdi RM, Bocquet JF, Chhor K, Pommier C (1992) Solubility and Decomposition Studies on Metal Chelates in Supercritical Fluids for Ceramic Precursor Powders Synthesis. J of Supercritical Fluids 4:55

Warzinski RP, Lee CH, Holder GD (1992) Supercritical-Fluid Solubilization of Catalyst Precursors: The Solubility and Phase Behavior Molybdenum Hexacarbonyl in Supercritical Carbon Dioxide and Application to the Direct Liquefaction of Coal. J of Supercritical Fluids 5:60

Killilea WR, Swallow KC, Hong GT (1992) The Fate of Nitrogen in Supercritical-Water Oxidation. J of Supercritical Fluids 5:72

De Haan AB, De Graauw J (1992) Separation of Alkanes and Aromatics with Supercritical C2H6, CO2, CCIF3, and CHF3. Sep Sci and Technol 27:43

Subject Index

Printing: Mercedesdruck, Berlin
Binding: Buchbinderei Lüderitz & Bauer, Berlin